Figure 2 Reactan...
(Always use correspon...

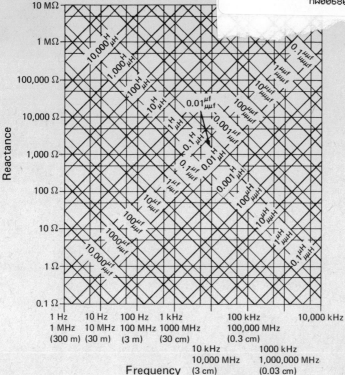

Reactance

Frequency

1 Hz	10 Hz	100 Hz	1 kHz	100 kHz	10,000 kHz
1 MHz	10 MHz	100 MHz	1000 MHz	100,000 MHz	
(300 m)	(30 m)	(3 m)	(30 cm)	(0.3 cm)	

10 kHz   1000 kHz
10,000 MHz   1,000,000 MHz
(3 cm)   (0.03 cm)

The accompanying chart is used to find:

1. The reactance of a given inductance at a given frequency.

2. The reactance of a given capacitance at a given frequency.

3. The resonant frequency of a given inductance and capacitance.

In order to facilitate the determination of magnitude of the quantities involved to two or three significant figures, the chart is divided into two parts. Figure 2 is the complete chart to be used for rough calculations. Figure 1, which is a single decade of Figure 2 enlarged approximately seven times, is to be used where the significant two or three figures are to be determined.

To Find Reactance

Enter the charts vertically from the bottom (frequency) and along the lines slanting upward to the left (inductance) or to the right (capacitance). Corresponding scales (upper or lower) must be used throughout. Project horizontally to the left from the intersection and read reactance value.

To Find Resonant Frequency

Enter the slanting lines for the given inductance and capacitance. Project downward from their intersection and read resonant frequency from the bottom scale. Corresponding scales (upper or lower) must be used throughout.

*Figures 1 and 2 are from General Radio Company Reactance Chart.

ELECTRONICS HANDBOOK

CLYDE HERRICK

San Jose City College

A Goodyear Applied Technical Book

Goodyear Publishing Company, Inc.
Pacific Palisades, California

Library of Congress Cataloging in Publication Data

Herrick, Clyde N.
   Electronics handbook.

   Includes index.
   1. Electronics—Handbooks, manuals, etc. I. Title.
TK7825.H46   621.381$'$02$'$02$'$ 74-15621
ISBN 0-87620-266-0

Copyright © 1975 by Goodyear Publishing Company, Inc.,
Pacific Palisades, California

Library of Congress Catalog Card Number: 74-15621

ISBN: 0-87620-266-0

Current Printing (last digit):
10 9 8 7 6 5 4 3 2 1

Y-2660-2

Printed in the United States of America

CONTENTS

HOW TO USE THIS HANDBOOK

In order to facilitate information retrieval, this handbook is divided into functional sections, subdivided into specific topics, and presented in graphical format. In turn, a sought equation, fact, or relation can be quickly located. An index is also provided to assist the reader. Graphical solutions are often important time-savers for the practical engineer and technician. Just as a slide rule eliminates much of the drudgery in numerical calculations, so do graphical solutions and universal charts save time and effort in algebraic operations. Sometimes a complex network can be completely solved graphically and an answer that is sufficiently accurate for engineering work obtained in a small fraction of the time required for routine algebraic analysis.

The following examples illustrate how to locate a specific equation, fact, or relation quickly.

1. Given the power and the resistance values in a dc circuit, find the voltage drop across the resistor.

This is an example of a basic law. Under Section 2, Basic Laws and Data, we note the topic Power Law: dc Circuits. In turn, we find the equation

$$E = \sqrt{PR}$$

We also observe this equation in the accompanying diagram, Figure 6.

2. Given the power, current, and power-factor values, find the impedance of an ac circuit.

This is another example of a basic law. Under Section 2, Basic Laws and Data, we note the topic Power Law: ac Circuits. In turn, we find the equation

$$Z = \frac{P}{I^2 \cos \theta}$$

We also observe this equation in the accompanying diagram, Figure 7.

3. Given the resistance and capacitance values of an RC integrating circuit, find the frequency response of the circuit.

This is an example of a variational characteristic of a circuit. Under Section 3, Variational Characteristics of Circuits, we note the topic Output Voltage Variation versus Frequency, RC Circuit. In turn, we find the universal charts in Figure 62, which show percentage of maximum output versus $\omega \, RC$ units.

4. Given the network shown below, reduce the network to its equivalent circuit at a given operating frequency.

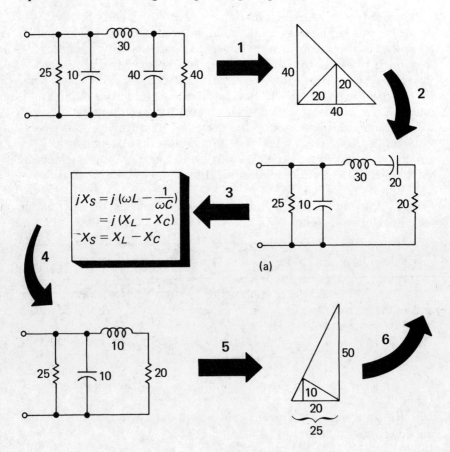

(a)

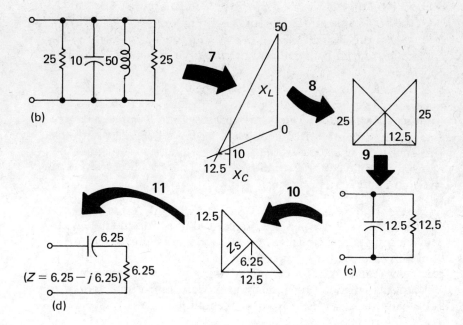

This is an example of the graphical solution of a network, using the graphical operations shown in this handbook. Note that we start with the reactances and resistances of the components at the stipulated operating frequency. Step 1 is shown in Figure 32. The resulting series RC values are drawn in Step 2. In the Step 3, the series inductive and capacitive values are subtracted ($30 - 20 = 10\ \Omega$, inductive). The resulting simplified circuit is drawn in Step 4. In Step 5, the series inductive reactance and resistance are converted into their parallel equivalent, as shown in Figure 32. In Step 6, the resulting parallel circuit is drawn. In Step 7, the parallel inductive and capacitive reactances are changed into their equivalent capacitive reactance, as shown in Figure 27. In Step 8, the parallel resistances are changed into their equivalent resistance, as shown in Figure 10. This gives us the equivalent parallel circuit for the original network. In turn, Step 9 converts the equivalent parallel circuit into its equivalent series circuit as shown in Figure 32.

5. Given an IF transformer, measure the value of the mutual inductance between primary and secondary.

This is an example of coupled–circuit characteristics. Under Section 12, Coupled Circuits, we note the topic Mutual Inductance, with associated Figure 142. In turn, we find the necessary test connections for inductance measurements that yield the value of the mutual inductance.

6. Given the cutoff frequency and load resistance for a constant-k low–pass filter section, find the required L and C values.

This is an example of a conventional filter problem. Under Section 13, Filter Networks, we are guided to Figure 147, which gives the applicable equations for the required L and C values.

7. Given the R and C values in a symmetrical two-section integrating circuit, find the relative output voltage at the end of two time constants.

This is an example of a transient response problem. Under Section 4, Transient Characteristics of Circuits, we note the topic Transient Response of Two- and Three Section Symmetrical RC Integrating Circuits, with accompanying Figure 69. This universal time-constant chart shows that the relative output voltage at the end of two time constants will be approximately 50 percent of the maximum output amplitude.

The foregoing examples are typical and illustrate the ease with which a sought equation, fact, or relation can be located in this handbook. In case the Contents is not a sufficiently precise guide in a particular situation, refer to the index if necessary. This handbook will save the engineer, technician, or student much time and effort in his daily tasks if it is utilized as exemplified above.

SECTION 1

RESISTOR COLOR CODE

Unless resistors are marked with the resistor value, they may have 3, 4, or 5 color bands to indicate the resistor's value and tolerance. The meaning of the bands for the four systems is as follows. In each case the first significant figure is closest to the edge of the resistor.

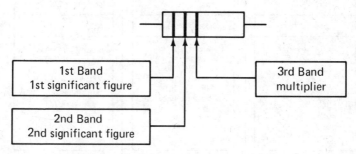

Figure 1.   3-band code for ± 20 percent resistors only

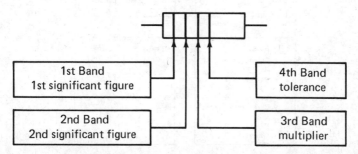

Figure 2.   4-band code for resistors with ± 1 percent to ± 10 percent tolerance

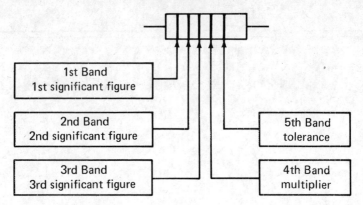

Figure 3. 5-band code for resistors with ± 1 percent to ± 10 percent tolerance

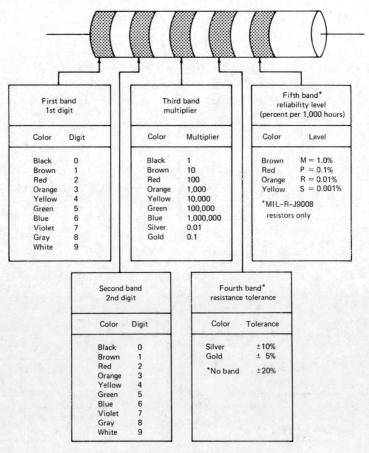

Figure 4. 5-band code for resistors with ± 1 percent to ± 10 percent tolerance (fifth band reliability level)

SECTION 2
BASIC LAWS AND DATA

Section 2 presents Ohm's law and the power laws for dc and ac circuits, resistance, capacitance, inductance, and reactance equations, impedance equations, characteristics of series and parallel circuits, phase angles of impedances, and power factors of resistive, inductive, capacitive, and impedance circuits.

OHM'S LAW: DC CIRCUITS

Figure 5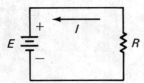

Basic equations:
$$I = \frac{E}{R}$$
$$E = IR$$
$$R = \frac{E}{I}$$

where I = current in amperes
E = voltage in volts
R = resistance in ohms

POWER LAW: DC CIRCUITS

Figure 6

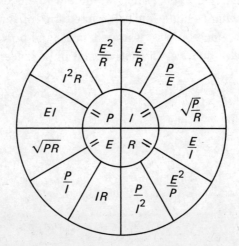

Basic equations:
$$P = EI = I^2R = \frac{E^2}{R}$$
$$I = \frac{P}{E} = \sqrt{\frac{P}{R}}$$
$$E = \frac{P}{I} = \sqrt{PR}$$
$$R = \frac{P}{I^2} = \frac{E^2}{P}$$

where P = power in watts

4

OHM'S LAW: AC CIRCUITS

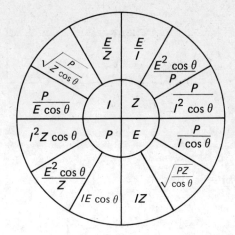

Figure 7

Basic equations:

$$I = \frac{E}{Z}$$
$$E = IZ$$
$$Z = \frac{E}{I}$$

where I = current in amperes
E = voltage across Z
Z = impedance in ohms

POWER LAW: AC CIRCUITS

Basic equations:

$$P = IE \cos \theta = I^2 Z \cos \theta = \frac{E^2 \cos \theta}{Z}$$
$$I = \frac{P}{E \cos \theta} = \sqrt{\frac{P}{Z \cos \theta}}$$
$$E = \frac{P}{I \cos \theta} = \sqrt{\frac{PZ}{\cos \theta}}$$
$$Z = \frac{E^2 \cos \theta}{P} = \frac{P}{I^2 \cos \theta}$$

where P = power in watts
θ = power-factor angle

5

RESISTANCE, CAPACITANCE, INDUCTANCE, AND REACTANCE EQUATIONS

Figure 8

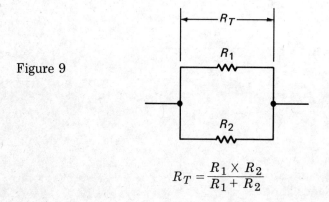

Resistors in series:

$$R_T = R_1 + R_2 + R_3 + \cdots$$

Two resistors in parallel:

Figure 9

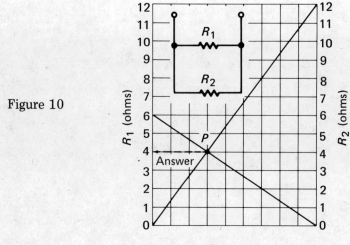

$$R_T = \frac{R_1 \times R_2}{R_1 + R_2}$$

Graphical solution for two resistors in parallel:

Figure 10

6

More than two resistors in parallel:

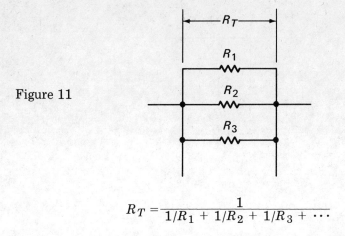

Figure 11

$$R_T = \frac{1}{1/R_1 + 1/R_2 + 1/R_3 + \cdots}$$

Graphical solution for more than two resistors in parallel:

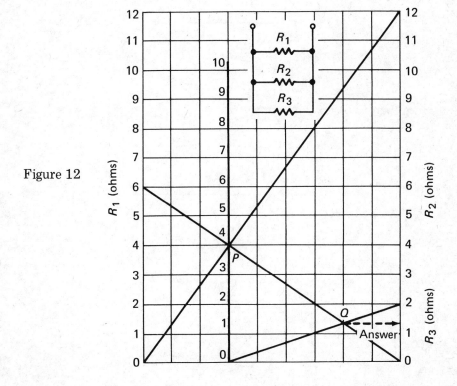

Figure 12

Capacitors in parallel:

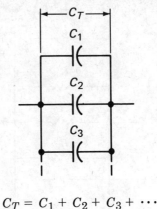

Figure 13

$$C_T = C_1 + C_2 + C_3 + \cdots$$

Two capacitors in series:

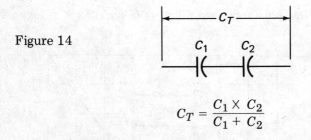

Figure 14

$$C_T = \frac{C_1 \times C_2}{C_1 + C_2}$$

More than two capacitors in series:

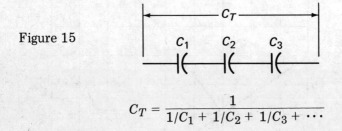

Figure 15

$$C_T = \frac{1}{1/C_1 + 1/C_2 + 1/C_3 + \cdots}$$

Inductors in series (with no mutual inductance):

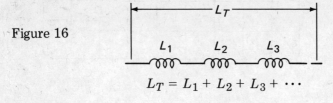

Figure 16

$$L_T = L_1 + L_2 + L_3 + \cdots$$

Two inductors in parallel (with no mutual inductance):

Figure 17

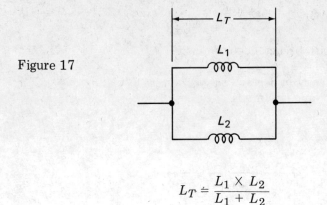

$$L_T \doteq \frac{L_1 \times L_2}{L_1 + L_2}$$

More than two inductors in parallel (with no mutual inductance):

Figure 18

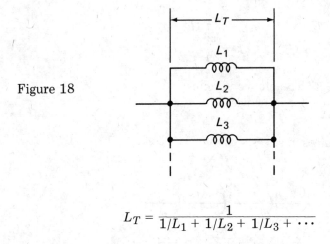

$$L_T = \frac{1}{1/L_1 + 1/L_2 + 1/L_3 + \cdots}$$

Capacitive reactance:

Figure 19

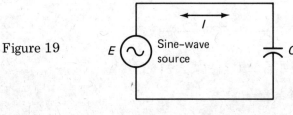

$$X_c = \frac{1}{2\pi fC}$$

$$I = \frac{E}{X_c}$$

where X_c = capacitive reactance in ohms
π = 3.1416
f = frequency in hertz
C = capacitance in farads

Inductive reactance:

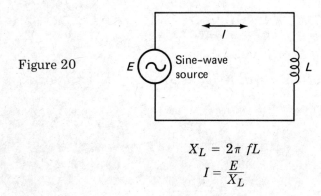

Figure 20

$$X_L = 2\pi fL$$

$$I = \frac{E}{X_L}$$

where X_L = inductive reactance in ohms
L = inductance in henries

Capacitive reactances in series:

Figure 21

$$X_T = X_{C1} + X_{C2} + X_{C3} + \cdots$$

Inductive reactances in series:

Figure 22

$$X_T = X_{L1} + X_{L2} + X_{L3} + \cdots$$

Capacitive reactances in parallel:

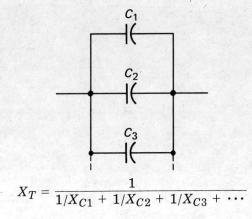

Figure 23

$$X_T = \frac{1}{1/X_{C1} + 1/X_{C2} + 1/X_{C3} + \cdots}$$

Inductive reactances in parallel:

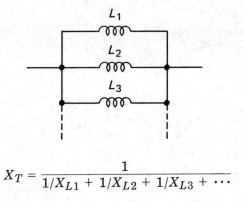

Figure 24

$$X_T = \frac{1}{1/X_{L1} + 1/X_{L2} + 1/X_{L3} + \cdots}$$

Inductive and capacitive reactances in series:

Figure 25

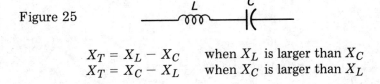

$$X_T = X_L - X_C \quad \text{when } X_L \text{ is larger than } X_C$$
$$X_T = X_C - X_L \quad \text{when } X_C \text{ is larger than } X_L$$

Inductive and capacitive reactances in parallel:

Figure 26

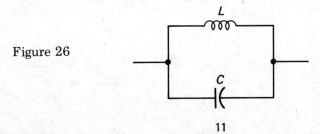

$$X_T = \frac{X_L X_C}{X_L - X_C} \qquad \text{when } X_L \text{ is larger than } X_C$$

$$X_T = \frac{X_C X_L}{X_C - X_L} \qquad \text{when } X_C \text{ is larger than } X_L$$

Graphical solution for inductive and capacitive reactances in parallel:

Figure 27

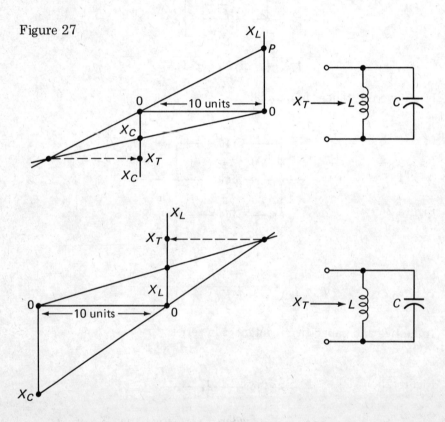

Resonant frequency for inductive and capacitive reactances:

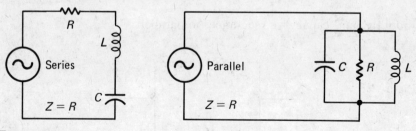

Figure 28

$$f_r = \frac{1}{2\pi \sqrt{LC}}$$

where f_r = resonant frequency in hertz
 L = inductance in henries
 C = capacitance in farads

Reactance of series LC circuit at resonance equals zero.
Reactance of parallel LC circuit at resonance equals infinity.

IMPEDANCE EQUATIONS

Relation of sides in a right triangle:

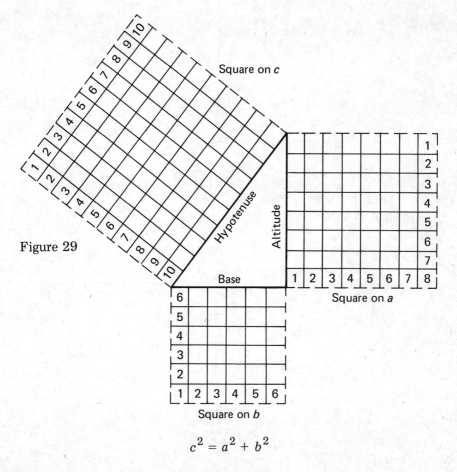

Figure 29

$$c^2 = a^2 + b^2$$

13

Impedance of a series RC circuit:

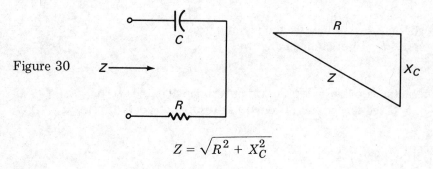

Figure 30

$$Z = \sqrt{R^2 + X_C^2}$$

where Z = impedance in ohms

Impedance of a series RL circuit:

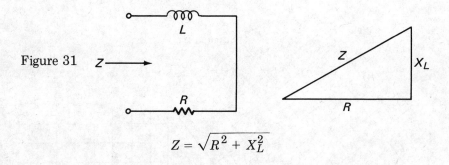

Figure 31

$$Z = \sqrt{R^2 + X_L^2}$$

Impedance of a parallel RC circuit:

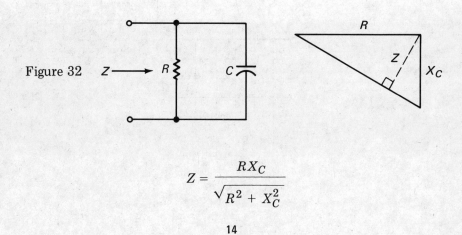

Figure 32

$$Z = \frac{RX_C}{\sqrt{R^2 + X_C^2}}$$

Impedance of a parallel RL circuit:

Figure 33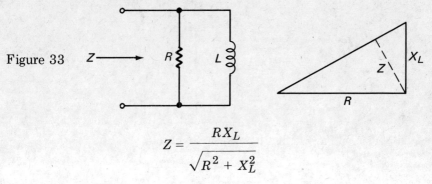

$$Z = \frac{RX_L}{\sqrt{R^2 + X_L^2}}$$

Equivalent series and parallel RC circuits:

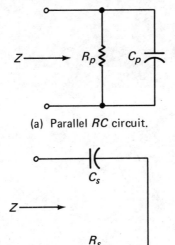

(a) Parallel RC circuit.

Figure 34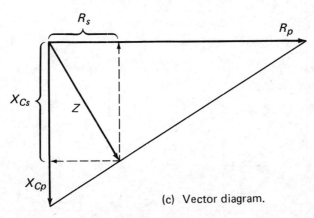

(b) Equivalent series circuit.

(c) Vector diagram.

15

Equivalent series and parallel RL circuits:

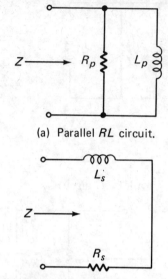

(a) Parallel RL circuit.

Figure 35

(b) Equivalent series RL circuit.

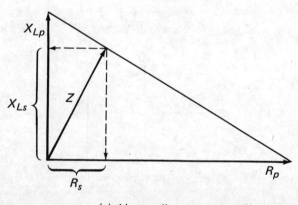

(c) Vector diagram.

Impedance of inductance and series resistance in parallel with resistance:

Figure 36

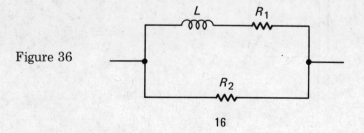

16

$$Z = R_2 \sqrt{\frac{R_1^2 + X_L^2}{(R_1 - R_2)^2 + X_L^2}}$$

Impedance of inductance, capacitance, and resistance in parallel:

Figure 37

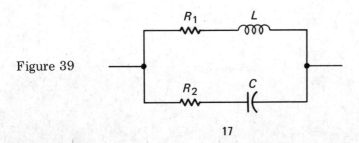

$$Z = \frac{R X_L X_C}{\sqrt{X_L^2 X_C^2 + R^2 (X_L - X_C)^2}}$$

Impedance of inductance and series resistance in parallel with capacitance:

Figure 38

$$Z = X_C \sqrt{\frac{R^2 + X_L^2}{R^2 + (X_L - X_C)^2}}$$

Impedance of capacitance and series resistance in parallel with inductance and series resistance:

Figure 39

$$Z = \sqrt{\frac{(R_1^2 + X_L^2)(R_2^2 + X_C^2)}{(R_1 + R_2)^2 + (X_L - X_C)^2}}$$

Impedance at resonance of inductance and series resistance in parallel with capacitance:

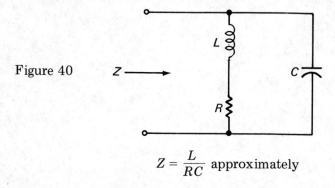

Figure 40

$$Z = \frac{L}{RC} \text{ approximately}$$

Impedance at resonance of series inductance, capacitance, and resistance:

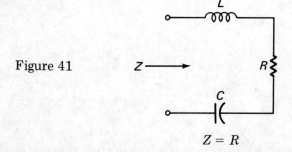

Figure 41

$$Z = R$$

CHARACTERISTICS OF SERIES- AND PARALLEL-RESONANT CIRCUITS

Quantity	Series Circuit	Parallel Circuit
At resonance: Reactance $(X_L - X_C)$	Zero; because $X_L = X_C$	Zero; because nonenergy currents are equal
Resonant frequency	$\dfrac{1}{2\pi\sqrt{LC}}$	$\dfrac{1}{2\pi\sqrt{LC}}$
Impedance	Minimum; $Z = R$	Maximum; $Z = \dfrac{L}{CR}$, approx.
I_{line}	Maximum	Minimum value
I_L	I_{line}	$Q \times I_{line}$
I_C	I_{line}	$Q \times I_{line}$
E_L	$Q \times E_{line}$	E_{line}
E_C	$Q \times E_{line}$	E_{line}
Phase angle between E_{line} and I_{line}	$0°$	$0°$
Angle between E_L and E_C	$180°$	$0°$
Angle between I_L and I_C	$0°$	$180°$
Desired value of Q	10 or more	10 or more
Desired value of R	Low	Low
Highest selectivity	High Q, low R, high $\dfrac{L}{C}$	High Q, low R
When f is greater than f_o: Reactance	Inductive	Capacitive
Phase angle between I_{line} and E_{line}	Lagging current	Leading current
When f is less than f_o: Reactance	Capacitive	Inductive
Phase angle between I_{line} and E_{line}	Leading current	Lagging current

Figure 42

PHASE ANGLES OF IMPEDANCES

Phase angle of a series LR circuit:

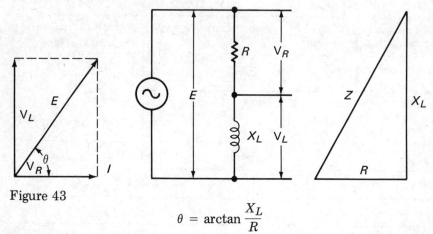

Figure 43

$$\theta = \arctan \frac{X_L}{R}$$

Phase angle of a series RC circuit:

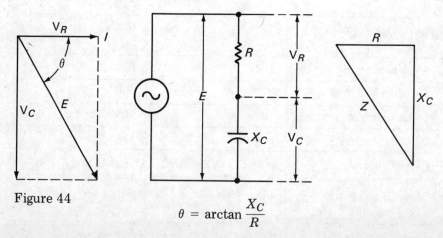

Figure 44

$$\theta = \arctan \frac{X_C}{R}$$

Phase angle of a parallel LR circuit:

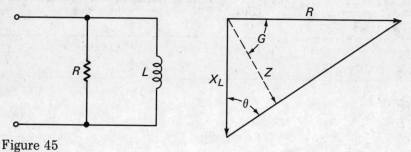

Figure 45

20

$$\theta = \arctan \frac{R}{X_L}$$

Phase angle of a parallel RC circuit:

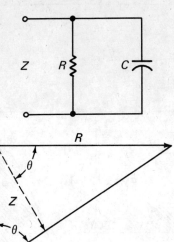

Figure 46

$$\theta = \arctan \frac{R}{X_C}$$

Phase angle of a series RLC circuit:

Figure 47

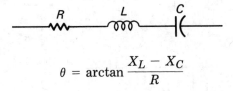

$$\theta = \arctan \frac{X_L - X_C}{R}$$

Phase angle of a parallel RLC circuit:

Figure 48

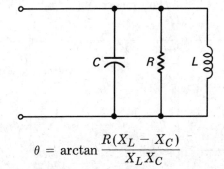

$$\theta = \arctan \frac{R(X_L - X_C)}{X_L X_C}$$

21

POWER FACTORS OF R, L, C, AND Z CIRCUITS

Power factor of a resistive circuit:

Figure 49

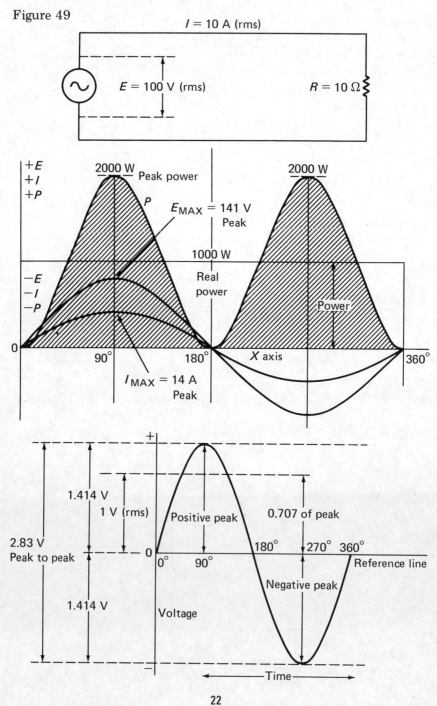

Current and voltage go through zero at the same instant; therefore the phase angle between current and voltage is zero. The power factor is equal to cos θ. Since the cosine of zero degrees is 1, the power factor is equal to 1.

$$P = EI \cos \theta = EI$$

In the example of Figure 49, the voltage and current have rms values of 100 V and 10 A. In turn, the value of the real power is 1000 W; the value of the peak power is 2000 W.

Power factor of a pure inductive circuit:

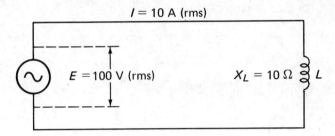

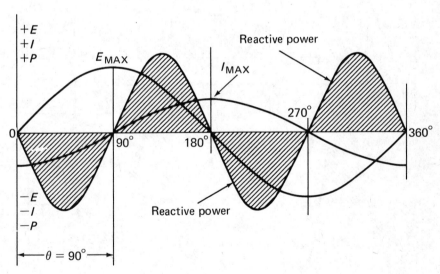

Figure 50

Current lags the voltage by 90°. Since the cosine of 90° is zero, the power factor is equal to zero.

$$P = EI \cos \theta = EI \times 0 = 0$$

23

In the example of Figure 50, the value of the real power is zero. On the other hand, the value of the reactive power is 1000 VARS.

$$\text{VARS} = EI \sin \theta$$

Reactive power (volt-amperes reactive) does no physical work but merely surges back and forth in the circuit.

In the example of Figure 50, the value of the apparent power is 1000 VA.

$$\text{Volt-Amperes} = EI$$

Power factor of a pure capacitive circuit:

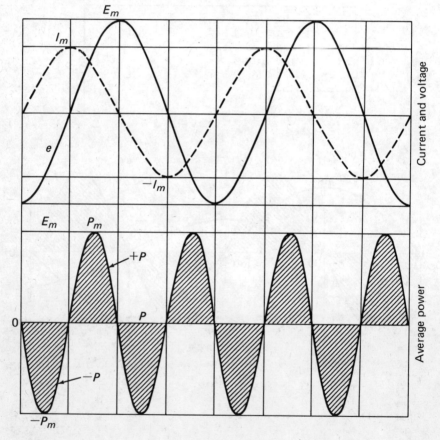

Figure 51

Current leads the voltage by 90°. The power factor is equal to zero, and all the power in the circuit is reactive power.

Power factor of an RL circuit:

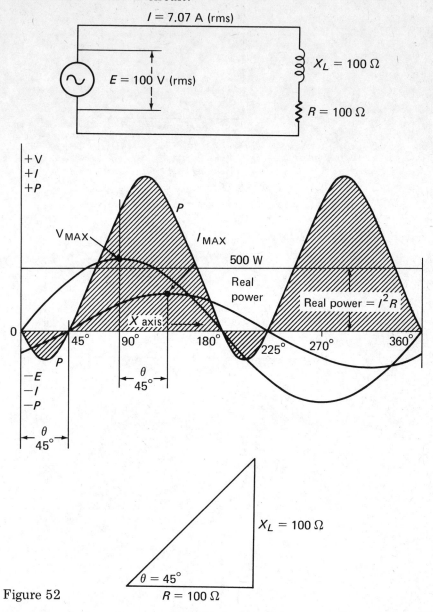

Figure 52

Current lags the voltage by 45°. Since the cosine of 45° is 0.707, the power factor is equal to 0.707. In turn, the real power is equal to 500 W. The sine of 45° is 0.707, and the reactive power is equal to 500 W. There are 707 VA in the circuit. The three power values combine in a right triangle, as shown in Figure 53, page 26.

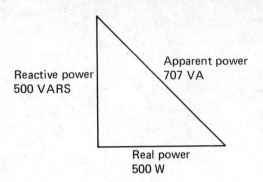

Figure 53

$$P = EI \cos \theta = I^2 R$$
$$\text{VARS} = EI \sin \theta$$
$$\text{Volt–Amperes} = EI$$

Power factor of an RC circuit:

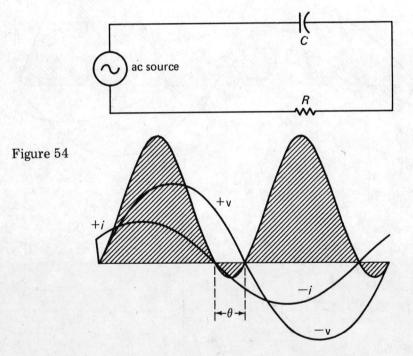

Figure 54

Current leads the voltage by the angle θ. Otherwise the power relations are the same as in an inductive circuit.

$$P = EI \cos \theta$$
$$\text{VARS} = EI \sin \theta$$
$$\text{Volt–Amperes} = EI$$

Power factor of RLC series circuit:

Subtract the smaller reactance from the larger reactance and proceed with the equivalent RC or RL circuit. If the difference of the reactances is zero, the circuit is resonant and the power factor is zero; the equivalent circuit becomes the resistance that is present, as shown in Figure 55.

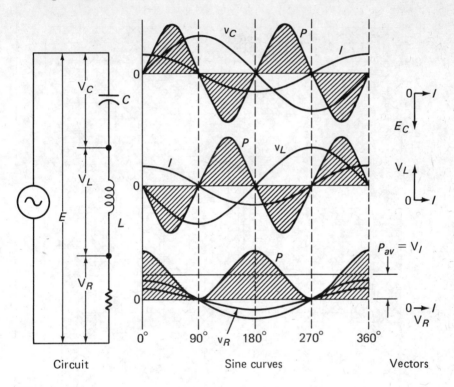

Circuit   Sine curves   Vectors

Figure 55

Reactance chart:

Follow chart directions to find the reactance value of an inductance or capacitance or to find the resonant frequency of an *LC* combination.

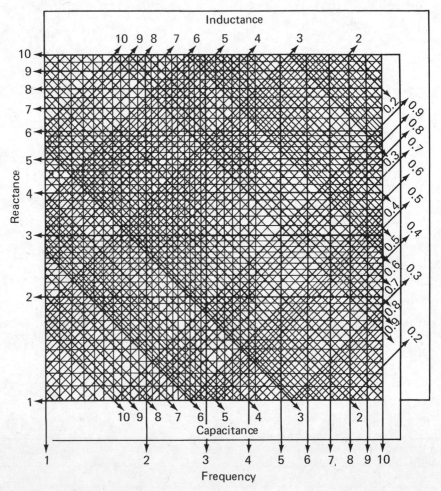

Figure 56. Reactance chart

After obtaining an approximate answer from Figure 57, proceed to find a more precise answer from Figure 56, if desired. The decimal point must be determined from Figure 57.

Examples: The reactance of a 0.00012–Hy inductance at 900 kHz is 737 Ω. A capacitance of 90 pF will have 250 Ω of reactance at 7 MHz. The resonant frequency of 159 pF and 159 μH is 1 MHz.

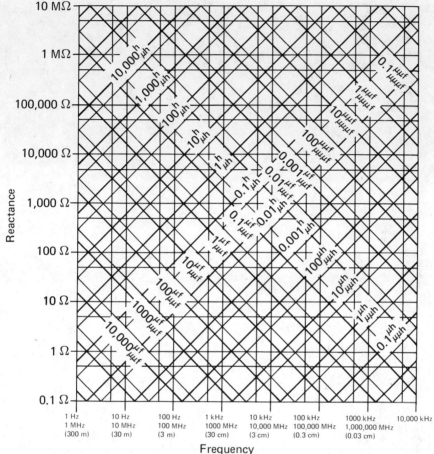

Figure 57. Reactance chart, cont'd.

Figure 57 may be used to find the reactance of an inductance at a given frequency, the reactance of a capacitance at a given frequency, or the resonant frequency of an *LC* combination. The chart has two parts; Figure 57 is a complete chart used for approximate answers. Figure 56 is a single decade which has been enlarged. Proceed by entering the charts vertically from the bottom ¬nd move along the lines slanting upward to the left (inductance lines), or along the lines slanting to the right (capacitance lines). The corresponding scales (upper or lower) must be used consistently. To find a reactance value, project horizontally to the left from the intersection. To find a resonant frequency, enter the slanting lines corresponding to the given inductance and capacitance values. Then project downward from the intersection and read the resonant frequency value on the bottom scale. Note that corresponding scales must be used consistently.

29

Summary of Ohm's law and power law equations:

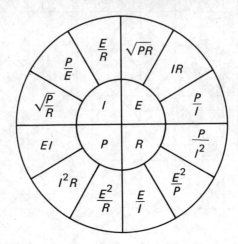

Figure 58

SECTION 3

VARIATIONAL CHARACTERISTICS OF CIRCUITS

Section 3 presents current variation versus resistance, power variation versus resistance and current, power variation versus voltage and current, output voltage variation versus frequency for RC circuits, output voltage variation versus frequency for RL circuits, output phase variation versus frequency for RC differentiating circuits, output phase variation versus frequency for RC integrating circuits, output voltage and phase variation for RLC series circuits, and output voltage and phase variation for RLC parallel circuits.

CURRENT VARIATION VERSUS RESISTANCE

Figure 59

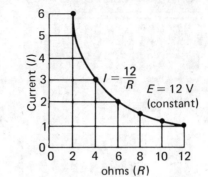

POWER VARIATION VERSUS RESISTANCE AND CURRENT

Figure 60

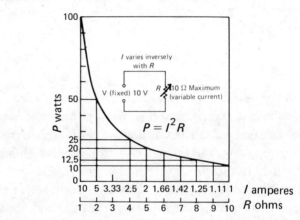

POWER VARIATION VERSUS VOLTAGE AND CURRENT

Figure 61

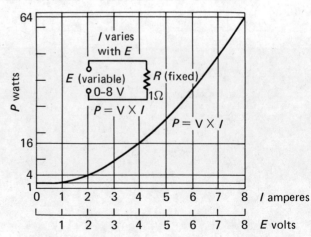

OUTPUT VOLTAGE VARIATION VERSUS FREQUENCY, RC CIRCUIT

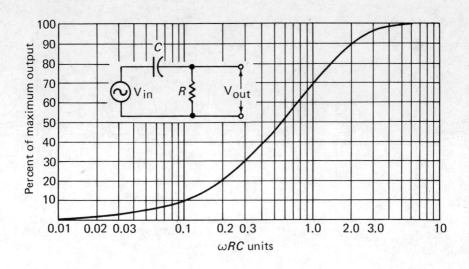

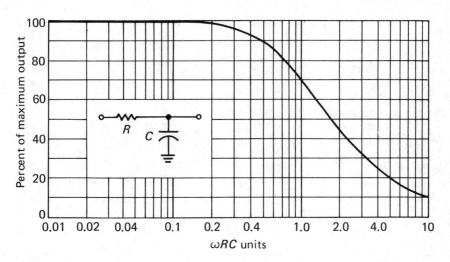

Figure 62

OUTPUT VOLTAGE VARIATION VERSUS FREQUENCY, *RL* CIRCUIT

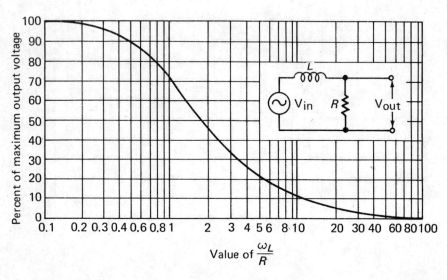

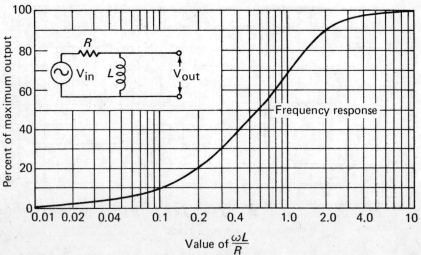

Figure 63

OUTPUT PHASE VARIATION VERSUS FREQUENCY, RC DIFFERENTIATING CIRCUIT

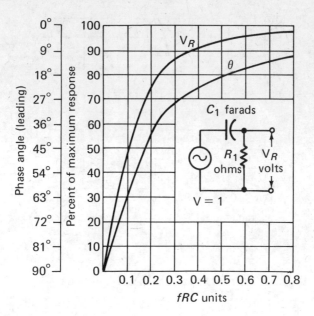

Figure 64

OUTPUT PHASE VARIATION VERSUS FREQUENCY, RC INTEGRATING CIRCUIT

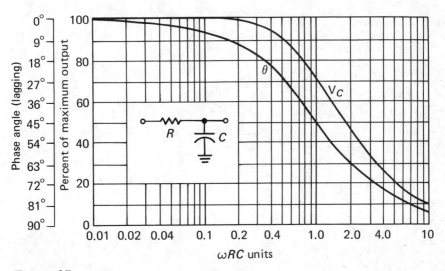

Figure 65

OUTPUT VOLTAGE AND PHASE VARIATION, *RLC* SERIES CIRCUIT

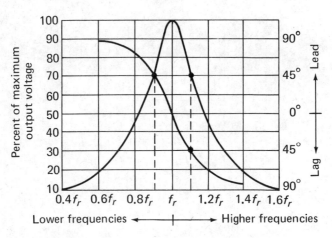

Figure 66

OUTPUT VOLTAGE AND PHASE VARIATION, *RLC* PARALLEL CIRCUIT

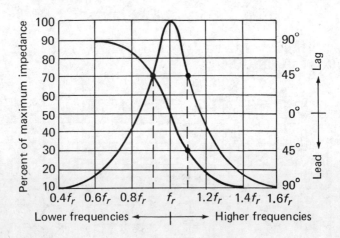

Figure 67

SECTION 4

TRANSIENT CHARACTERISTICS OF CIRCUITS

Section 4 presents the transient response of series RC circuits, transient response of two- and three-section symmetrical integrating circuits, transient response of two-section symmetrical RC integrating circuits with device isolation, comparative transient responses of RC integrating circuits with various time constants, transient response of a two-section symmetrical RC differentiator, transient response of a two-section symmetrical differentiator with device isolation, transient response of a series RL circuit, equivalent RC and RL/RC two-section symmetrical integrating circuits, transient response of an LC resonant circuit, and transient response of a critically damped LCR circuit.

TRANSIENT RESPONSE OF A SERIES *RC* CIRCUIT

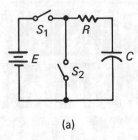

(a)

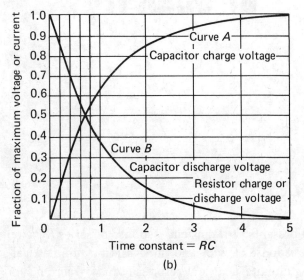

(b)

Figure 68. Charge and discharge curves for series *RC* circuit.
(a) Circuit.
(b) Universal time–constant chart.

Time constant $T = RC$

where T is in seconds, R in ohms, and C in farads.

$$e = E\varepsilon^{-t/RC}$$
$$e = E(1 - \varepsilon^{-t/RC})$$

where $\varepsilon = 2.718$

TRANSIENT RESPONSE OF TWO- AND THREE–SECTION SYMMETRICAL RC INTEGRATING CIRCUITS

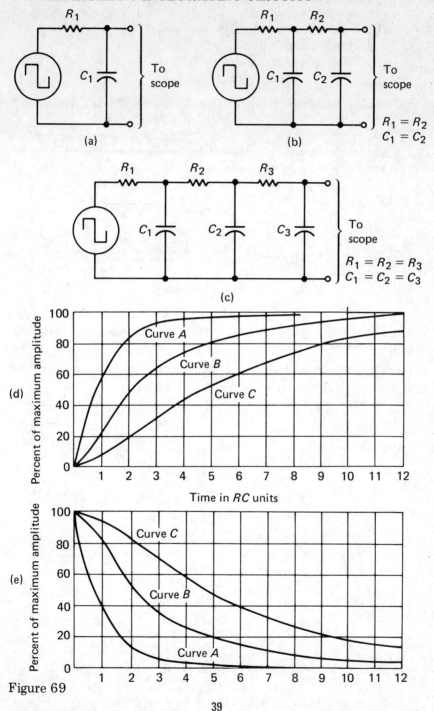

Figure 69

Charge and discharge curves for two- and three-section symmetrical integrating circuits (see Figure 69, page 39.)
(a) One-section circuit.
(b) Two-section circuit.
(c) Three-section circuit.
(d) Leading–edge transient responses.
(e) Trailing–edge transient responses.

TRANSIENT RESPONSE OF TWO–SECTION SYMMETRICAL RC INTEGRATING CIRCUIT WITH DEVICE ISOLATION

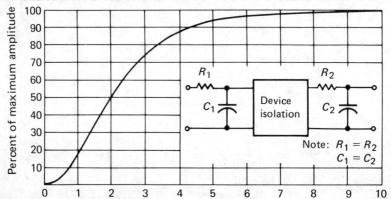

Figure 70. Charge curve for two–section RC integrating circuit with device isolation.

COMPARATIVE TRANSIENT RESPONSES OF RC INTEGRATING CIRCUITS WITH VARIOUS TIME CONSTANTS

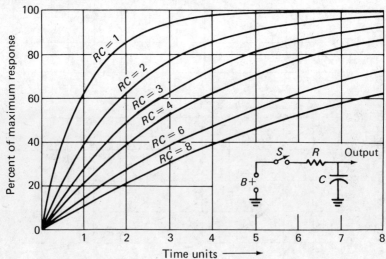

Figure 71. Comparative integrator waveforms for various time constants.

TRANSIENT RESPONSE OF A TWO-SECTION SYMMETRICAL RC DIFFERENTIATOR

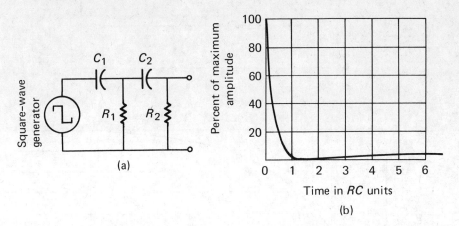

Figure 72. Transient response of a two-section symmetrical RC differentiator.
(a) Circuit.
(b) Universal time–constant chart.

TRANSIENT RESPONSE OF A TWO-SECTION SYMMETRICAL DIFFERENTIATOR WITH DEVICE ISOLATION

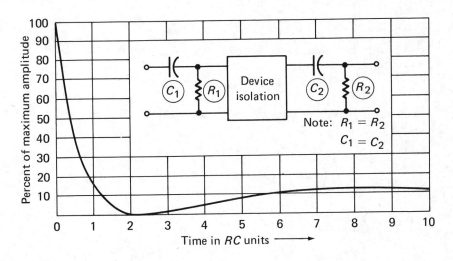

Figure 73. Transient response of a two-section symmetrical RC differentiator with device isolation.

TRANSIENT RESPONSE OF A SERIES *RL* CIRCUIT

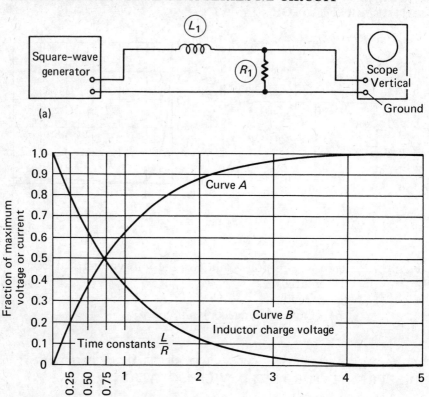

(a)

(b)

Figure 74. Transient response of a series *RL* circuit.
    (a) Circuit.
    (b) Universal *L/R* time-constant chart.

Note that distributed capacitance of the inductor may modify the ideal response.

EQUIVALENT RC AND RL/RC TWO–SECTION SYMMETRICAL INTEGRATING CIRCUITS

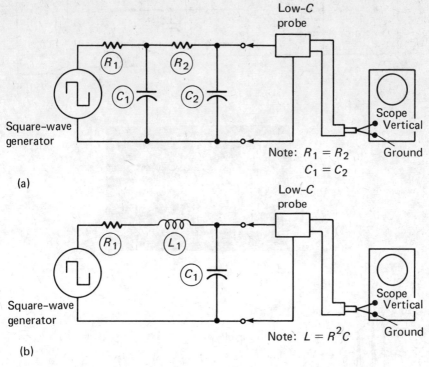

Note: $R_1 = R_2$
$C_1 = C_2$

(a)

Note: $L = R^2 C$

(b)

Figure 75. Equivalent RC and RL/RC two–section symmetrical integrating circuits.
(a) RC form.
(b) RL/RC form.

Note that distributed capacitance of the inductor may modify the ideal response.

TRANSIENT RESPONSE OF AN *LC* RESONANT CIRCUIT

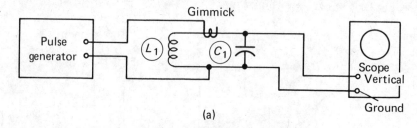

(a)

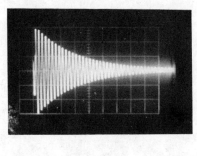

(b)

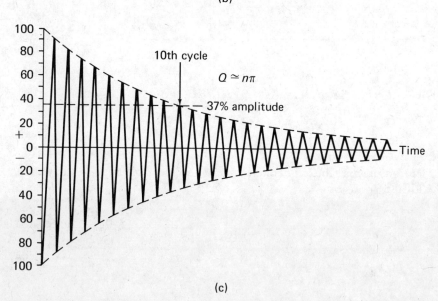

(c)

Figure 76. Transient response of *LC* resonant circuit.
   (a) Test connections.
   (b) Typical ringing waveform.
   (c) Determination of approximate *Q* value.

The Q value of the inductor at the ringing frequency is approximately $n\pi$, where n is the number of cycles (peaks) in the ringing waveform from its maximum point to its 37 percent of maximum point. Thus the Q value in the example is equal to approximately 31.4. From a practical point of view, the Q value is determined by the coil-winding resistance.

TRANSIENT RESPONSE OF A CRITICALLY DAMPED *LCR* CIRCUIT

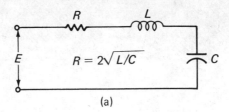

(a)

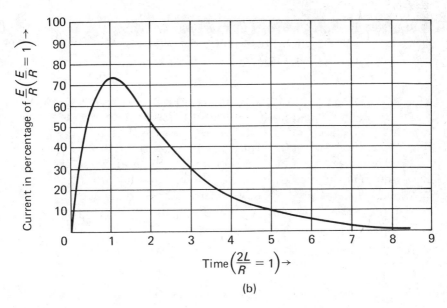

(b)

Figure 77. Transient response of a critically damped *LCR* circuit.
     (a) Circuit.
     (b) Transient waveform.

Critical damping occurs when $R = 2\sqrt{L/C}$. Any smaller value of R causes underdamping and a ringing transient response.

POWER AND ENERGY IN *RC* AND *RL* TRANSIENT CIRCUITS

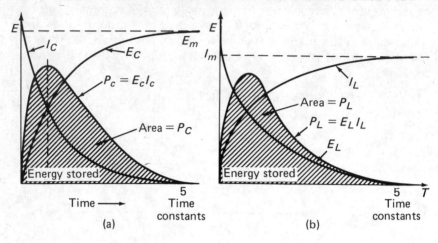

Figure 78. Power and energy in an *RC* or *RL* transient circuit.
    (a) *RC* circuit response.
    (b) *RL* circuit response.

Energy is stored progressively by the reactor in an *RC* or *RL* circuit when voltage is applied at the input of the circuit. Energy storage is essentially complete at the end of five time constants. Energy is measured in watt–seconds. The rate of energy storage is the power in the circuit. Power is measured in watts. The power value rises to a maximum at the crossover time of the voltage and current curves and then decreases progressively to zero. The amount of energy stored in the reactor is proportional to the area under the power curve.

SECTION 5

CHARACTERISTICS OF RESONANT CIRCUITS

Section 5 presents the basic characteristics of series-resonant circuits, parallel-resonant circuits, and a summary of basic formulas containing a Q-value term.

SERIES–RESONANT CIRCUIT

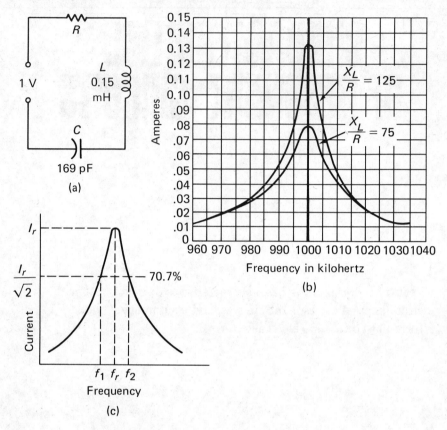

Figure 79.   Example of series resonance.
(a) Circuit.
(b) Frequency response for two values of Q.
(c) Definition of bandwidth.

$$\text{Resonant frequency } f_r = \frac{1}{2\pi\sqrt{LC}}$$

Impedance at resonance is equal to the circuit resistance R.

Q value is equal to X_L/R at the resonant frequency.

Voltage magnification at resonance is equal to Q times the applied voltage and appears across both the inductor and the capacitor.

Bandwidth at resonance is equal to f_r/Q; the bandwidth is also equal to the number of hertz between the half–power points (70.7% of maximum points on the frequency response curve).

48

PARALLEL-RESONANT CIRCUIT

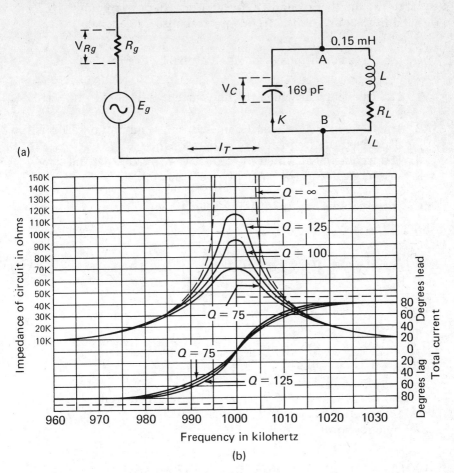

(a)

(b)

Figure 80. Example of parallel resonance.
    (a) Circuit
    (b) Frequency response and phase characteristic for
    four values of Q.

$$\text{Resonant frequency } f_r = \frac{1}{2\pi \sqrt{LC}}$$

Impedance at resonance is approximately equal to L/RC ohms.

Q value is equal to Z_r/X_L at the resonant frequency, where Z_r is the impedance at resonance.

Current magnification at resonance is equal to Q times the line current and flows in both the inductive and the capacitive branches.

49

Bandwidth at resonance is equal to f_r/Q; the bandwidth is also equal to the number of hertz between the half-power points (70.7% of maximum points on the frequency response curve).

There are three conditions of parallel resonance in Figure 80, as follows:

1. The frequency at which the line current is in phase with the line voltage.
2. The frequency at which the line current is minimum. This is also called the frequency of anti-resonance.
3. The frequency at which the inductive reactance has the same absolute value as the capacitive reactance.

These three parallel-resonant frequencies are given by the following equations:

(1)
$$f_r = \frac{1}{2\pi} \sqrt{\frac{1}{LC} - \frac{R^2}{L^2}}$$

(2)
$$f_r = \frac{1}{2\pi} \sqrt{\frac{1}{LC} - \frac{R^4 C}{2L^3}}$$

(3)
$$f_r = \frac{1}{2\pi} \sqrt{\frac{1}{LC}}$$

Note that as the coil Q value increases, the resonant frequencies stated by equations (1) and (2) approach the resonant frequency stated by equation (3).

SUMMARY OF BASIC FORMULAS CONTAINING A Q-VALUE TERM

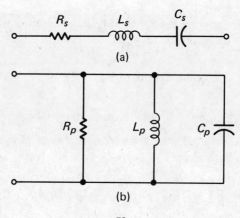

Figure 81

(a)

(b)

Figure 81. Series and parallel LCR circuits.
  (a) L_s, C_s, and R_s in series circuit.
  (b) L_p, C_p, and R_p in parallel circuit.

$$Q = \frac{\omega L_s}{R_s} = \frac{1}{\omega C_s R_s} = \frac{R_p}{\omega L_p} = R_p \omega C_p = \frac{\sqrt{L_s/C_s}}{R_s} = \frac{R_p}{\sqrt{L_p/C_p}}$$

General Formulas	Formulas for Q Greater than 10	Formulas for Q Less than 0.1
$R_s = \dfrac{R_p}{1 + Q^2}$	$R_s \simeq \dfrac{R_p}{Q_2}$	$R_s \simeq R_p$
$X_s = X_p \dfrac{Q^2}{1 + Q^2}$	$X_s \simeq X_p$	$X_s \simeq X_p Q^2$
$L_s = L_p \dfrac{Q^2}{1 + Q^2}$	$L_s \simeq L_p$	$L_s \simeq L_p Q^2$
$C_s = C_p \dfrac{1 + Q^2}{Q^2}$	$C_s \simeq C_p$	$C_s \simeq \dfrac{C_p}{Q^2}$
$R_p = R_s (1 + Q^2)$	$R_p \simeq R_s Q^2$	$R_p \simeq R_s$
$X_p = X_s \dfrac{1 + Q^2}{Q^2}$	$X_p \simeq X_s$	$X_p \simeq \dfrac{X_s}{Q^2}$
$L_p = L_s \dfrac{1 + Q^2}{Q^2}$	$L_p \simeq L_s$	$L_p \simeq \dfrac{L_s}{Q^2}$
$C_p = C_s \dfrac{Q^2}{1 + Q^2}$	$C_p \simeq C_s$	$C_p \simeq C_s Q^2$
$B_L = \dfrac{1}{X_L}$	$B_L = \dfrac{1}{X_L}$	$B_L = \dfrac{1}{X_L}$
$B_c = \dfrac{1}{X_c}$	$B_L = \dfrac{1}{X_L}$	$B_L = \dfrac{1}{X_L}$
$Y = \sqrt{G^2 + B^2}$	$Y = \dfrac{1}{\sqrt{G^2 + B^2}}$	$Y = \sqrt{G^2 + B^2}$

SECTION 6
DECIBEL RELATIONS

Section 6 presents the principles of decibel measurements, a decibel table, and a list of decibel reference levels.

DECIBEL MEASUREMENTS

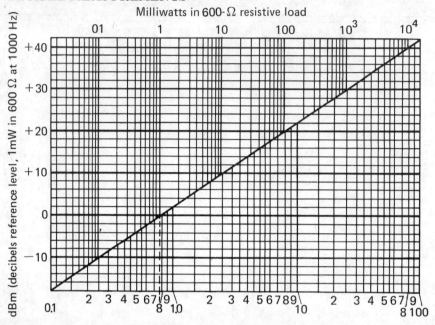

Figure 82. ac voltages corresponding to dBm values.

A dBm value is the number of dB above or below a reference level of 1 mW in 600 Ω. In turn, ac voltage values measured across a 600–Ω load correspond to dBm values. If the load has a resistance greater or less than 600 Ω, a load–correction factor must be utilized. A load–correction table is provided in Figure 83.

Resistive Load at 1000 Hz	dBm*
600	0
500	+0.8
300	+3.0
250	+3.8
150	+6.0
50	+10.8
15	+16.0
8	+18.8
3.2	+22.7

*dBm is the increment to be added algebraically to the dBm value read from the graph.

Figure 83. Load–correction table for dBm values.

The general formulas for decibel ratios are as follows:

$$\text{Bels} = \log P_2/P_1$$
$$\text{Decibels} = 10 \log P_2/P_1$$
$$\text{Decibels} = 20 \log V_2/V_1 \text{ (measured across equal loads)}$$
$$\text{Decibels} = 20 \log I_2/I_1 \text{ (measured across equal loads)}$$
$$\text{Decibels} = 20 \log V_2/V_1 + \log 10\, R_1/R_2$$
$$\text{Decibels} = 20 \log I_2/I_1 + R_2/R_1$$

If a dB meter has a reference level of zero dB equal to 1 mW in 600 Ω, it can be used to measure dB gain or loss across other values of load, as 1000 Ω, for example. If the input resistance of an amplifier is 1000 Ω, and its output resistance is 1000 Ω, the difference in dB readings across the input and output terminals will be the actual number of dB gain. On the other hand, if the input and output resistances are unequal, a correction factor must be applied, as shown in Figure 84. As an illustration, if the input resistance is 1000 Ω and the output resistance is 10 Ω, the resistance ratio is 100, and a correction factor of 20 dB must be added to the measured difference in dB readings.

DECIBEL TABLE

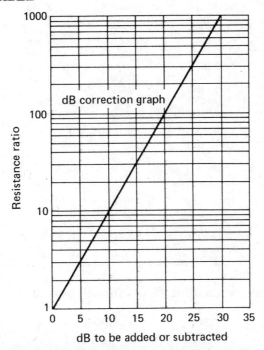

Figure 84. Correction graph for dB values measured across a pair of unequal load resistances.

Note: If an amplifier distorts the applied sine–input waveform, the output dB reading will be incorrect due to waveform error.

Power Ratio	Voltage Ratio	dB − + ← →	Voltage Ratio	Power Ratio
1.000	1.000	0	1.000	1.000
0.9772	0.9886	0.1	1.012	1.023
0.9550	0.9772	0.2	1.023	1.047
0.9333	0.9661	0.3	1.035	1.072
0.9120	0.9550	0.4	1.047	1.096
0.8913	0.9441	0.5	1.059	1.122
0.8710	0.9333	0.6	1.072	1.148
0.8511	0.9226	0.7	1.084	1.175
0.8318	0.9120	0.8	1.096	1.202
0.8128	0.9016	0.9	1.109	1.230
0.7943	0.8913	1.0	1.122	1.259
0.6310	0.7943	2.0	1.259	1.585
0.5012	0.7079	3.0	1.413	1.995
0.3981	0.6310	4.0	1.585	2.512
0.3162	0.5623	5.0	1.778	3.162
0.2512	0.5012	6.0	1.995	3.981
0.1995	0.4467	7.0	2.239	5.012
0.1585	0.3981	8.0	2.512	6.310
0.1259	0.3548	9.0	2.818	7.943
0.10000	0.3162	10.0	3.162	10.00
0.07943	0.2818	11.0	3.548	12.59
0.06310	0.2512	12.0	3.981	15.85
0.05012	0.2293	13.0	4.467	19.95
0.03981	0.1995	14.0	5.012	25.12
0.03162	0.1778	15.0	5.623	31.62
0.02512	0.1585	16.0	6.310	39.81
0.01995	0.1413	17.0	7.079	50.12
0.01585	0.1259	18.0	7.943	63.10
0.01259	0.1122	19.0	8.913	79.43
0.01000	0.1000	20.0	10.000	100.00
10^{-3}	3.162×10^{-2}	30.0	3.162×10	10^3
10^{-4}	10^{-2}	40.0	10^2	10^4
10^{-5}	3.162×10^{-3}	50.0	3.162×10^2	10^5
10^{-6}	10^{-3}	60.0	10^3	10^6
10^{-7}	3.162×10^{-4}	70.0	3.162×10^3	10^7
10^{-8}	10^{-4}	80.0	10^4	10^8
10^{-9}	3.162×10^{-5}	90.0	3.162×10^4	10^9
10^{-10}	10^{-5}	100.0	10^5	10^{10}

Figure 85. Tabulation of dB values versus power and voltage ratios.

DECIBEL REFERENCE LEVELS

Any convenient reference level may be defined as zero dB. Some of the common established reference levels are

dBk—1 kilowatt
dBm—1 milliwatt, 600 ohms
dBv—1 volt
dBw—1 watt
dBvg—voltage gain
dBrap—decibels above a reference acoustical power of 10^{-16} watt
VU—1 milliwatt, 600 ohms (complex waveforms varying in both amplitude and frequency)

SECTION 7

NONSINUSOIDAL WAVEFORM CHARACTERISTICS

Section 7 presents the rms values of common complex waveforms, the response of various voltmeters to complex waveforms, response of a meter movement to half-sine waves, the
effect of phase relations on complex waveforms, the harmonic
composition of common complex waveforms, and basic square-
wave distortions.

SECTION 7

rms VALUES OF COMMON COMPLEX WAVEFORMS

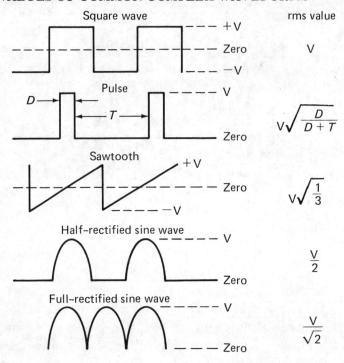

Waveform	rms value
Square wave	V
Pulse	$V\sqrt{\dfrac{D}{D+T}}$
Sawtooth	$V\sqrt{\dfrac{1}{3}}$
Half-rectified sine wave	$\dfrac{V}{2}$
Full-rectified sine wave	$\dfrac{V}{\sqrt{2}}$

Figure 86. rms values of common complex waveforms.

RESPONSE OF VARIOUS VOLTMETERS TO COMPLEX WAVEFORMS

Note: V = peak amplitude of wave.

Waveform	Meter Response	Scale Reading
Square wave	Peak	0.707 A
	Half-wave average	1.11 A
	Full-wave average	1.11 A
Sawtooth wave	Peak	0.707 A
	Half-wave average	0.555 A
	Full-wave average	0.555 A
Half-rectified sine wave	+Peak	0.707 A
	+Half-wave average	0.707 A
	Full-wave average	0.354 A
Full-rectified sine wave	Peak	0.707 A
	+Half-wave average	1.414 A
	Full-wave average	0.707 A

Figure 87. Response of various voltmeters to complex waveforms.

RESPONSE OF METER MOVEMENT TO HALF–SINE WAVES

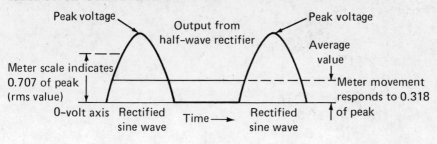

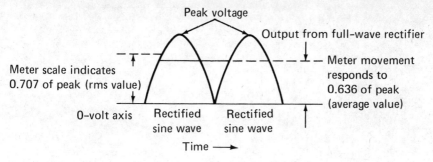

Figure 88. Response of meter movement to half–sine waves.

Note: A dc voltmeter reads 0.318 of the peak voltage for a half-rectified sine wave and reads 0.636 of the peak voltage for a full-rectified sine wave.

EFFECT OF PHASE RELATIONS ON COMPLEX WAVEFORM

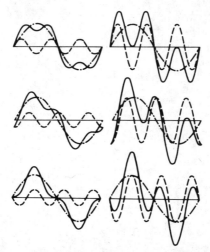

Figure 89. Examples of a fundamental and third harmonic in various phase relations.

SECTION 7

HARMONIC COMPOSITION OF COMMON COMPLEX WAVEFORMS

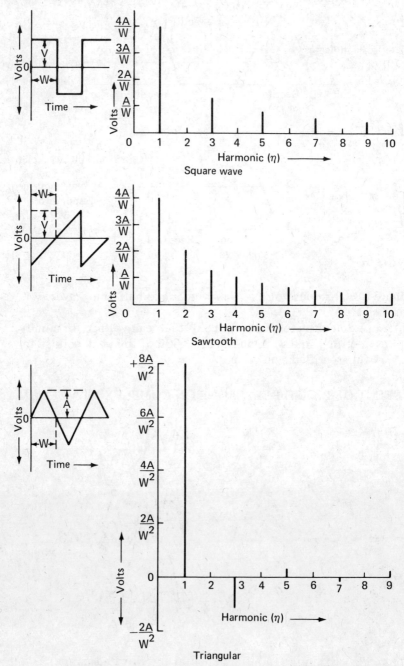

Figure 90. Harmonic composition of three complex waveforms.

62

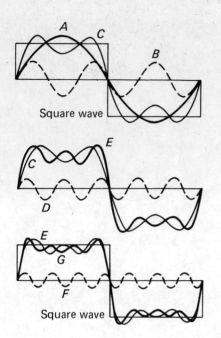

A. Fundamental
B. Third harmonic
C. Fundamental plus third harmonic
D. Fifth harmonic
E. Fundamental plus third and fifth harmonics
F. Seventh harmonic
G. Fundamental plus third, fifth, and seventh harmonics

Figure 90 (continued). Phase relations of fundamental and harmonics in a square wave.

BASIC SQUARE-WAVE DISTORTIONS

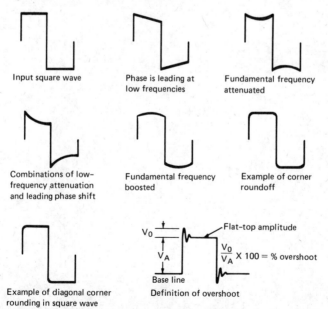

Input square wave

Phase is leading at low frequencies

Fundamental frequency attenuated

Combinations of low-frequency attenuation and leading phase shift

Fundamental frequency boosted

Example of corner roundoff

Example of diagonal corner rounding in square wave

Definition of overshoot

$$\frac{V_0}{V_A} \times 100 = \% \text{ overshoot}$$

Figure 91. Basic square-wave distortions.

63

SECTION 8

MODULATED WAVEFORM CHARACTERISTICS

Section 8 presents amplitude modulation versus linear mixing, amplitude modulation percentage, amplitude modulation with carrier suppression, single-sideband, amplitude-modulated waveforms, stereomultiplex, composite audio signals, frequency-modulated waveforms, and pulse-modulated waveforms.

AMPLITUDE MODULATION VERSUS LINEAR MIXING

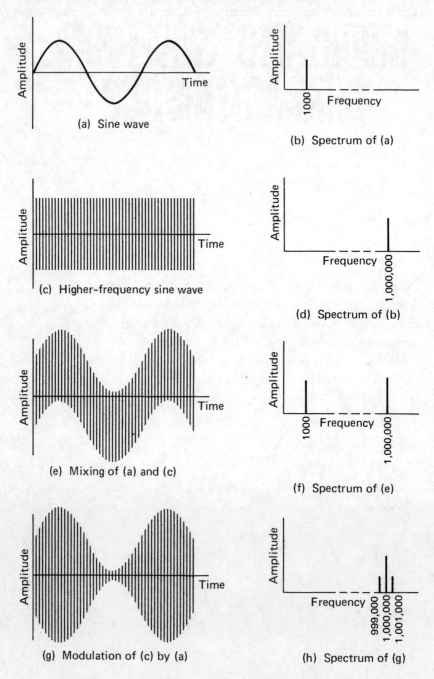

(a) Sine wave

(b) Spectrum of (a)

(c) Higher–frequency sine wave

(d) Spectrum of (b)

(e) Mixing of (a) and (c)

(f) Spectrum of (e)

(g) Modulation of (c) by (a)

(h) Spectrum of (g)

Figure 92. Examples of linear mixing and amplitude modulation.

AMPLITUDE MODULATION PERCENTAGE

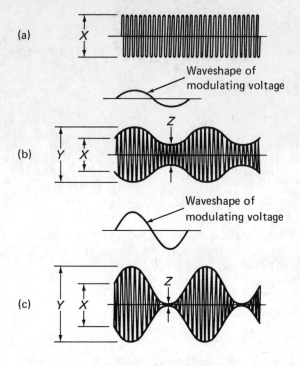

Figure 93. Amplitude modulation percentage.
(a) Unmodulated carrier; zero–percent modulation.
(b) Carrier modulated 50 percent.
(c) Carrier modulated 100 percent.

Percent modulation $= \dfrac{Y - X}{X} \times 100$ (upward modulation)

Percent modulation $= \dfrac{X - Z}{X} \times 100$ (downward modulation)

AMPLITUDE MODULATION WITH CARRIER SUPPRESSION

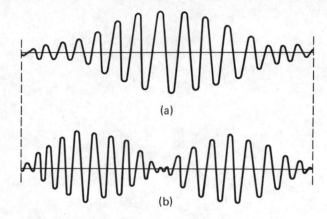

(a)

(b)

Figure 94. Amplitude modulation with carrier suppression.
(a) Carrier with both sidebands.
(b) Sidebands without carrier.

Note: The sidebands–without–carrier waveform has a double–frequency modulation envelope.

SINGLE–SIDEBAND, AMPLITUDE–MODULATED WAVEFORM

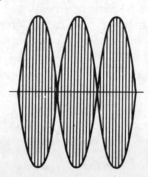

Figure 95. Single–sideband, 100 percent amplitude–modulated waveform.

Note: The single–sideband, 100 percent amplitude–modulated waveform has a modulation envelope that differs from the modulating waveshape. That is, when the modulating waveshape is sinusoidal, the modulation envelope consists of half–sine waves. However, the distortion is less for reduced percentages of modulation.

STEREOMULTIPLEX, COMPOSITE AUDIO SIGNAL

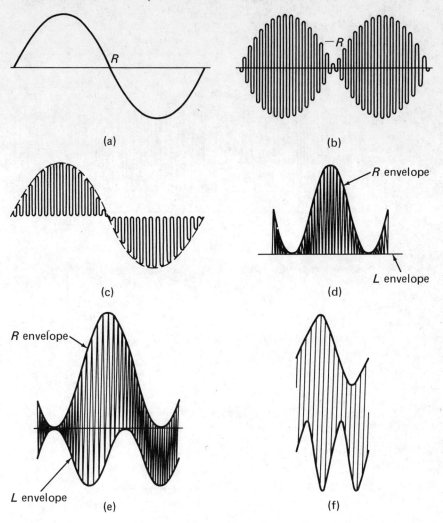

Figure 96. Stereomultiplex, composite audio signal.
   (a) Right–channel modulating signal.
   (b) Sidebands without 38-kHz subcarrier.
   (c) Right–channel modulating signal plus sidebands.
   (d) R channel energized; composite signal at receiver with
      subcarrier inserted.
   (e) Both R and L channels energized; composite signal at receiver
      with subcarrier inserted.
   (f) Both R and L channels energized; composite signal at receiver
      with subcarrier inserted at increased amplitude.

FREQUENCY-MODULATED WAVEFORM

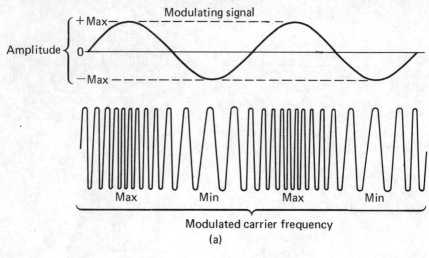

Modulated carrier frequency

(a)

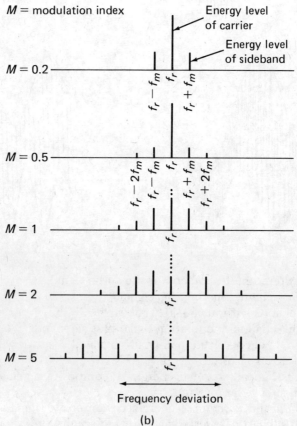

Figure 97

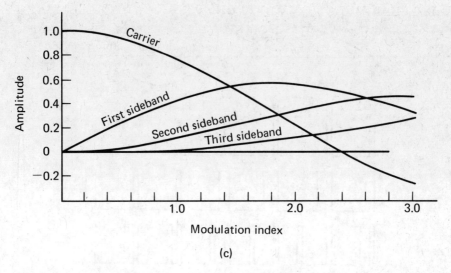

(c)

Figure 97 (continued). Frequency modulation characteristics.
      (a) Development of FM waveform.
      (b) Sideband frequencies in an FM waveform.
      (c) The amplitude of the pairs of sidebands varies with the modulation index.

Note: The modulation index is equal to f_d/f_m, where f_d is the frequency deviation of the carrier and f_m is the modulating frequency. The amount of deviation is proportional to the amplitude of the modulating frequency. The sideband frequency components are spaced apart by a number of hertz equal to the modulating frequency. When the modulation index exceeds 0.5, the carrier energy is significantly reduced. If the modulation index exceeds 2.2, the carrier amplitude goes through zero and then becomes reversed in phase or its amplitude changes from positive to negative.

PULSE MODULATION WAVEFORMS

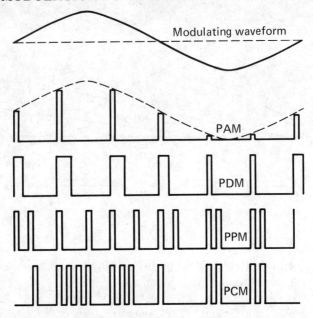

Figure 98. Basic types of pulse modulation.

The basic types of pulse modulation are pulse–amplitude modulation (PAM), pulse–duration modulation (PDM), pulse–position modulation (PPM), and pulse–code modulation (PCM).

SECTION 9

BASIC OSCILLOSCOPE DISPLAYS

Section 9 presents Lissajous figures, phase-angle measurement, audio-amplifier distortion checks, trapezoidal percentage-modulation waveforms, ac waveforms with dc components, oscilloscope calibration, stereomultiplex, composite audio signals, rise-time measurement, pulse-width measurement, and square-wave tilt measurement.

LISSAJOUS FIGURES

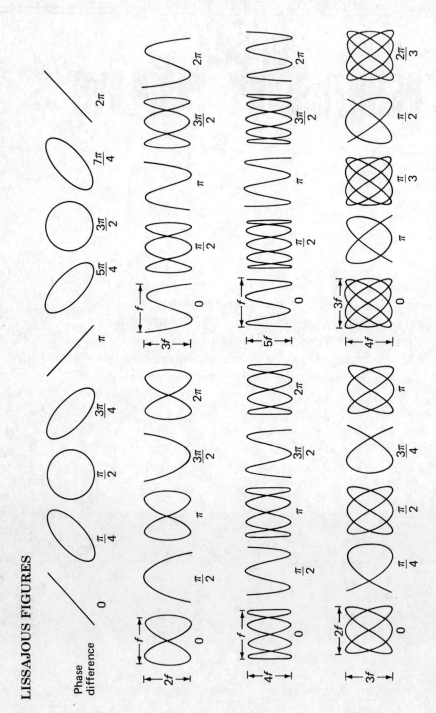

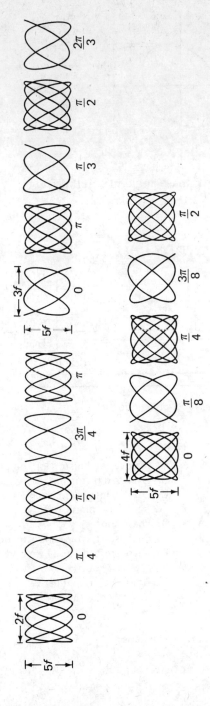

Figure 99. Basic Lissajous figures.

Lissajous figures are used to measure frequency and phase relations. One sine–wave voltage is applied to the vertical–input terminals of an oscilloscope, and another sine–wave voltage is applied to the horizontal–input terminals. Stationary patterns are displayed if the two frequencies are integrally related.

PHASE-ANGLE MEASUREMENT

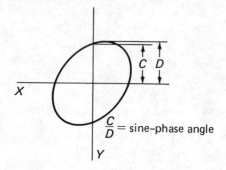

$\dfrac{C}{D}$ = sine-phase angle

Figure 100.   Measurement of phase angle in an elliptical
Lissajous figure.

AUDIO-AMPLIFIER DISTORTION CHECK

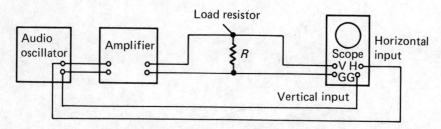

(a) Test connections.

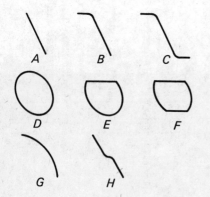

A. No overload distortion, no phase shift
B. Overload distortion, no phase shift
C. Driving into grid current, and past
cutoff, no phase shift
D. Phase shift
E. Phase shift, overload distortion
F. Phase shift, driving into grid current,
and past cutoff
G. Amplitude nonlinearity
H. Crossover distortion

(b) Pattern evaluations.

Figure 101. Audio-amplifier distortion check.

TRAPEZOIDAL PERCENTAGE-MODULATION WAVEFORMS

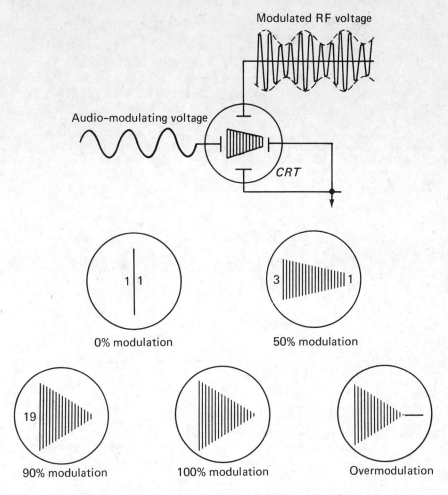

Figure 102. Trapezoidal percentage-modulation waveforms.

AC WAVEFORM WITH DC COMPONENT

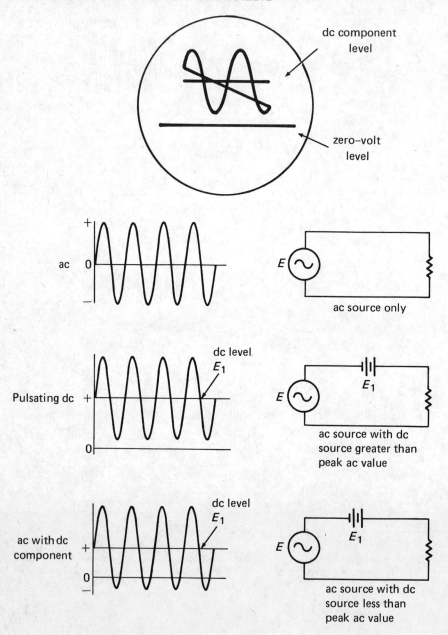

Figure 103. Examples of ac waveforms with dc components.

Note: The dc component level is displayed only if a dc oscilloscope is employed. An oscilloscope will reject the dc component.

OSCILLOSCOPE CALIBRATION

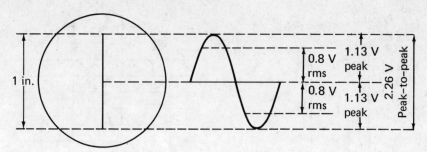

Figure 104. Calibration of oscilloscope for a deflection factor of
2.26 V peak-to-peak per inch.

STEREOMULTIPLEX, COMPOSITE AUDIO SIGNAL

Figure 105. Stereomultiplex, composite audio signal
with 19-kHz pilot subcarrier.

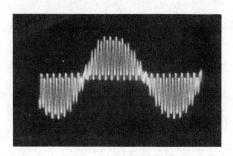

RISE–TIME MEASUREMENT

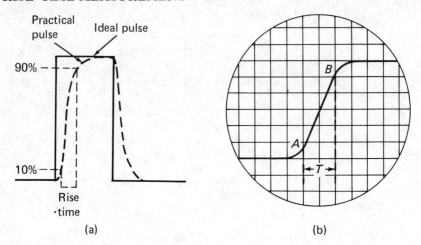

Figure 106. Rise–time measurement.
> (a) Rise–time interval.
> (b) Expansion of leading edge for convenient
> rise–time measurement.

The rise time is measured from the 10 percent to the 90 percent amplitude points on the leading edge of the waveform. Measurement is in time units.

PULSE–WIDTH MEASUREMENT

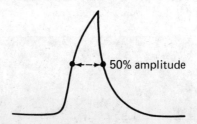

Figure 107. Pulse–width measurement.

Pulse width is measured between the 50 percent amplitude points on the leading and trailing edges of the waveform. Measurement is in time units.

SQUARE-WAVE TILT MEASUREMENT

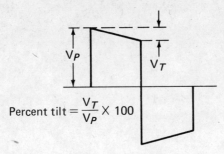

$$\text{Percent tilt} = \frac{V_T}{V_P} \times 100$$

Figure 108. Measurement of square-wave tilt.

Note: The low-frequency cutoff point of an amplifier is given by

$$f_c = \frac{2f(V_P - V_T)}{3(V_P + V_T)}$$

where
f_c = cutoff frequency
f = square-wave frequency
V_P and V_T = amplitudes shown in Figure 108

Conversely, the high-frequency cutoff point of an amplifier is given by

$$f_c = 0.35T$$

where f_c = frequency at which the response is down 3 dB
T = rise time of the amplifier

SECTION 10

BASIC TELEVISION WAVEFORMS

Section 10 presents the composite video waveform, RF frequency response curves, IF frequency response curves, video frequency response curves, color TV frequency response curves, NTSC color-bar signal with unadjusted values, NTSC color-bar signal with readjusted values, color-burst waveform, NTSC color-bar signal components with readjusted values, oscilloscope display of NTSC color-bar signal, keyed rainbow signal and color-bar pattern, chroma-demodulator output waveforms, phases of primary and complementary color signals, vectorgram display with unkeyed rainbow signal, ideal vectorgram with keyed rainbow signal, typical vectorgram displays, and the NTSC vectorgram pattern.

COMPOSITE VIDEO WAVEFORM

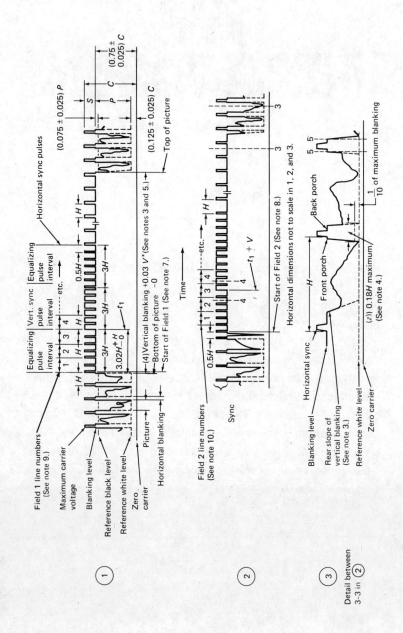

84

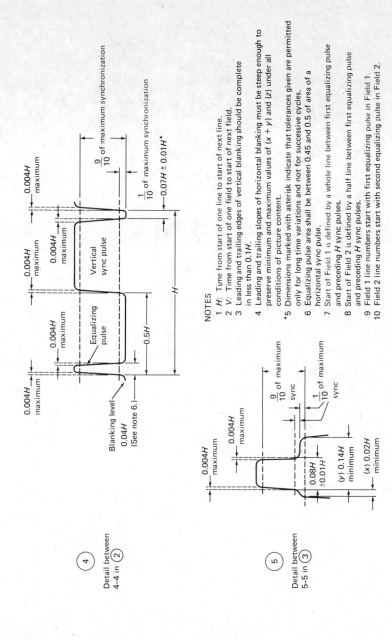

NOTES

1 H: Time from start of one line to start of next line.
2 V: Time from start of one field to start of next field.
3 Leading and trailing edges of vertical blanking should be complete in less than $0.1H$.
4 Leading and trailing slopes of horizontal blanking must be steep enough to preserve minimum and maximum values of $(x + y)$ and (z) under all conditions of picture content.
*5 Dimensions marked with asterisk indicate that tolerances given are permitted only for long time variations and not for successive cycles.
6 Equalizing pulse area shall be between 0.45 and 0.5 of area of a horizontal sync pulse.
7 Start of Field 1 is defined by a whole line between first equalizing pulse and preceding H sync pulses.
8 Start of Field 2 is defined by a half line between first equalizing pulse and preceding H sync pulses.
9 Field 1 line numbers start with first equalizing pulse in Field 1.
10 Field 2 line numbers start with second equalizing pulse in Field 2.

Figure 109. TV synchronizing waveform for monochrome transmission.

Source: Federal Communications Commission, Washington, D.C.

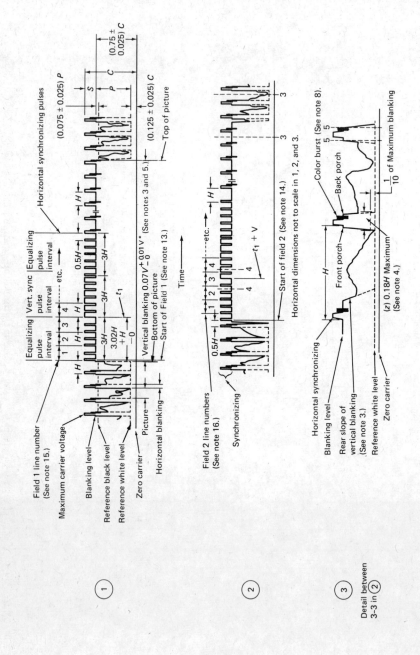

Field 1 line number (See note 15.)

Maximum carrier voltage

Blanking level

Reference black level

Reference white level

Zero carrier

Picture

Horizontal blanking

①

Horizontal synchronizing pulses

$(0.075 \pm 0.025)\,P$

$(0.75 \pm 0.025)\,C$

$(0.125 \pm 0.025)\,C$

Top of picture

Equalizing pulse interval | Vert. sync pulse interval | Equalizing pulse interval

Vertical blanking $0.07V^{+0.01V^*}_{-0}$ (See notes 3 and 5.)

Bottom of picture

Start of Field 1 (See note 13.)

Time

Field 2 line numbers (See note 16.)

Synchronizing

②

$t_1 + V$

Start of field 2 (See note 14.)

Horizontal dimensions not to scale in 1, 2, and 3.

③ Detail between 3-3 in ②

Color burst (See note 8).

Back porch

Horizontal synchronizing

Blanking level

Rear slope of vertical blanking (See note 3.)

Reference white level

Zero carrier

Front porch

H

$(z)\,0.18H$ Maximum (See note 4.)

$\frac{1}{10}$ of Maximum blanking

86

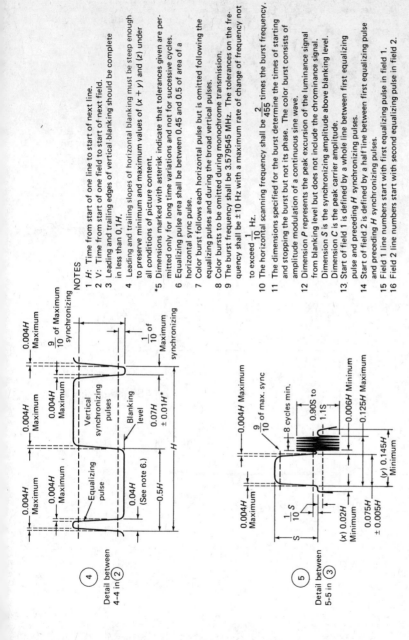

NOTES

1 H: Time from start of one line to start of next line.
2 V: Time from start of one field to start of next field.
3 Leading and trailing edges of vertical blanking should be complete in less than 0.1H.
4 Leading and trailing slopes of horizontal blanking must be steep enough to preserve minimum and maximum values of (x + y) and (z) under all conditions of picture content.
*5 Dimensions marked with asterisk indicate that tolerances given are permitted only for long time variations and not for successive cycles.
6 Equalizing pulse area shall be between 0.45 and 0.5 of area of a horizontal sync pulse.
7 Color burst follows each horizontal pulse but is omitted following the equalizing pulses and during the broad vertical pulses.
8 Color bursts to be omitted during monochrome transmission.
9 The burst frequency shall be 3.579545 MHz. The tolerances on the frequency shall be ±10 Hz with a maximum rate of change of frequency not to exceed $\frac{1}{10}$ Hz.
10 The horizontal scanning frequency shall be $\frac{2}{455}$ times the burst frequency.
11 The dimensions specified for the burst determine the times of starting and stopping the burst but not its phase. The color burst consists of amplitude modulation of a continuous sine wave.
12 Dimension P represents the peak excursion of the luminance signal from blanking level but does not include the chrominance signal.
    Dimension S is the synchronizing amplitude above blanking level.
    Dimension C is the peak carrier amplitude.
13 Start of field 1 is defined by a whole line between first equalizing pulse and preceding H synchronizing pulses.
14 Start of field 2 is defined by a half line between first equalizing pulse and preceding H synchronizing pulses.
15 Field 1 line numbers start with first equalizing pulse in field 1.
16 Field 2 line numbers start with second equalizing pulse in field 2.

Figure 110. TV synchronizing waveform for color transmission.

Source: Federal Communications Commission, Washington, D.C.

87

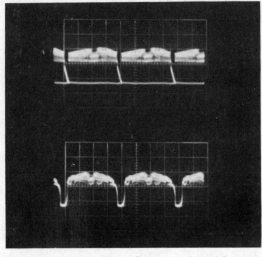

(a) Vertical-sync pulse
in composite signal

(b) Horizontal-sync pulse
in composite signal

Figure 111. Display of composite video waveform on oscilloscope screen (camera signal from test pattern).

RF FREQUENCY RESPONSE CURVE

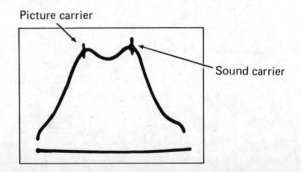

Figure 112. An RF frequency response curve.

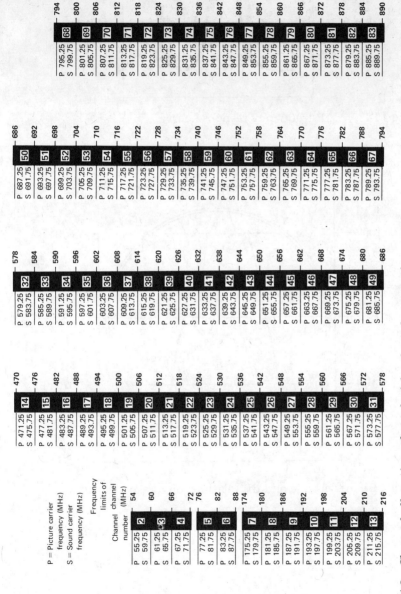

Figure 113. Channel allocations for television broadcast stations.

Source: Howard W. Sams & Co., Inc.

IF FREQUENCY RESPONSE CURVE

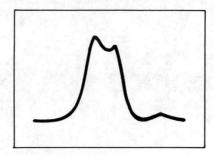

Figure 114. An IF frequency response curve.

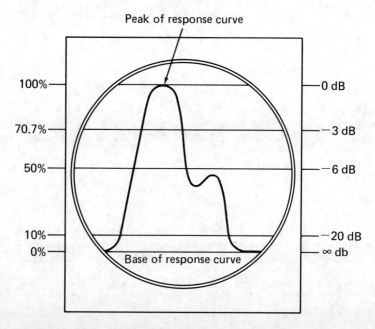

Figure 115. Amplitude and dB relations on a response curve.

VIDEO FREQUENCY RESPONSE CURVE

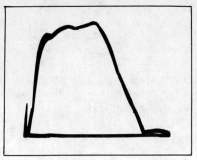

Figure 116. A video frequency response curve.

COLOR TV FREQUENCY RESPONSE CURVES

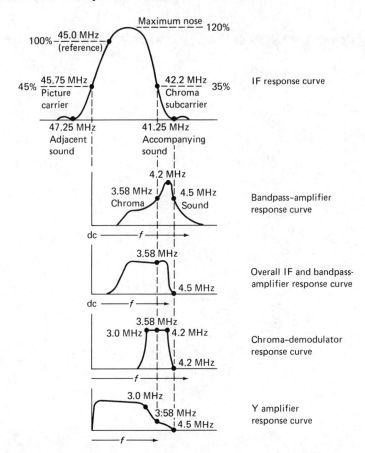

Figure 117. Color TV frequency response curves.

NTSC COLOR–BAR SIGNAL; UNADJUSTED VALUES

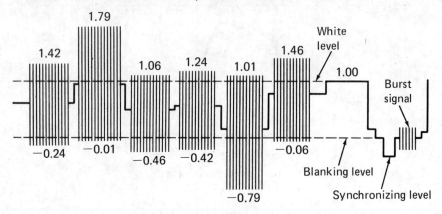

Figure 118. NTSC color–bar signal; unadjusted values.

NTSC COLOR–BAR SIGNAL; READJUSTED VALUES

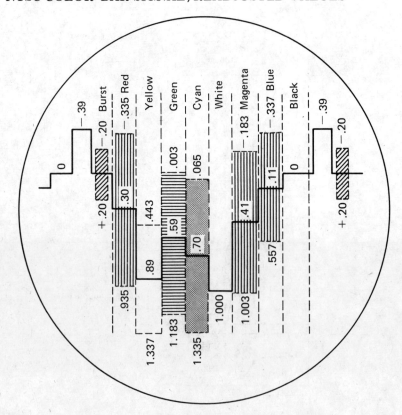

Figure 119. NTSC color–bar signal; readjusted values.

COLOR–BURST WAVEFORM

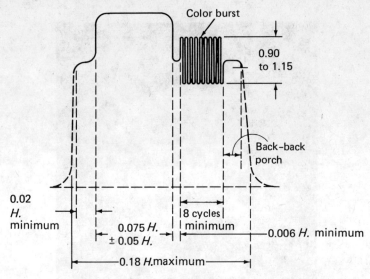

Figure 120. Color–burst waveform.

NTSC COLOR–BAR SIGNAL COMPONENTS; READJUSTED VALUES

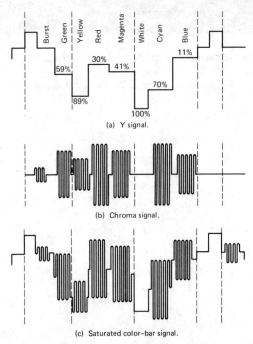

Figure 121. NTSC color–bar signal components; readjusted values.

93

OSCILLOSCOPE DISPLAY OF NTSC COLOR–BAR SIGNAL

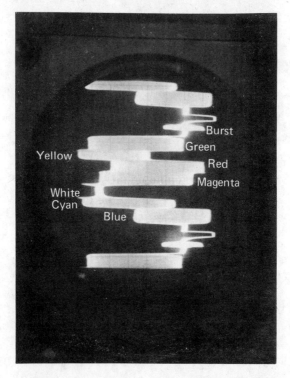

Figure 122. Oscilloscope display of NTSC color–bar signal.

Source: Howard W. Sams & Co., Inc.

KEYED RAINBOW SIGNAL AND COLOR-BAR PATTERN

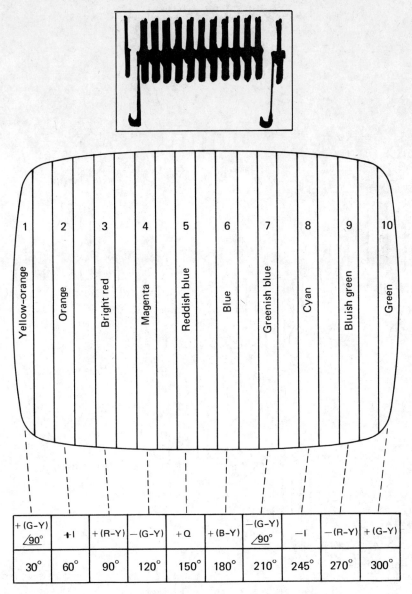

+ (G-Y) $\angle 90°$	+I	+ (R-Y)	− (G-Y)	+ Q	+ (B-Y)	− (G-Y) $\angle 90°$	−I	− (R-Y)	+ (G-Y)
30°	60°	90°	120°	150°	180°	210°	245°	270°	300°

Figure 123. Keyed rainbow signal and color-bar pattern.

CHROMA–DEMODULATOR OUTPUT WAVEFORMS

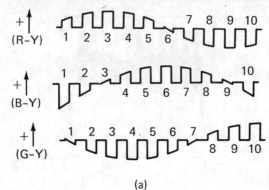

(a)

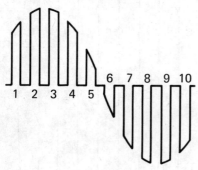

1. Red waveform.

2. Blue waveform.

3. Green waveform.

(b)

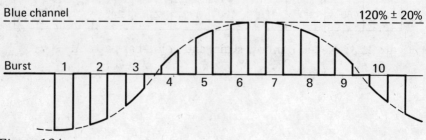

Blue channel

120% ± 20%

Burst

Figure 124

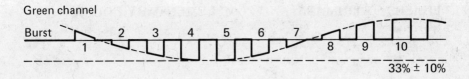

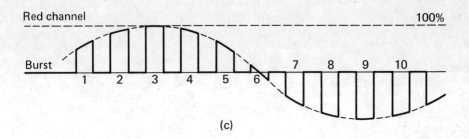

(c)

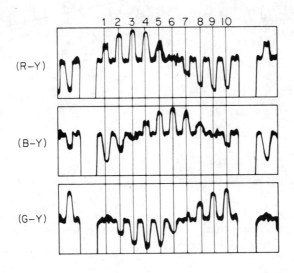

(d)

Figure 124 (continued). Chroma–demodulator output waveforms.

    (a) R-Y/B-Y/G-Y demodulators; ideal.
    (b) RGB demodulators; ideal.
    (c) *XZ* (105°) demodulators; ideal.
    (d) Typical R-Y/B-Y/G-Y waveforms.

Source: Howard W. Sams & Co., Inc.

97

PHASES OF PRIMARY AND COMPLEMENTARY COLOR SIGNALS

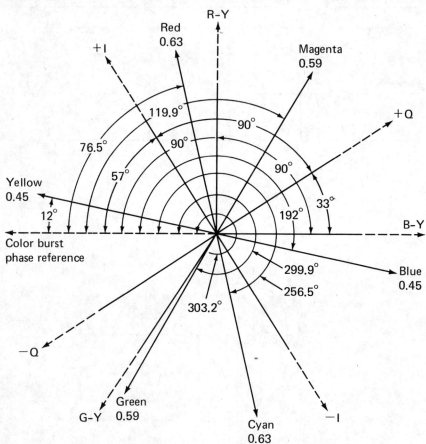

Figure 125. Phases of the primary, complementary, and basic color-difference signals.

VECTORGRAM DISPLAY WITH UNKEYED RAINBOW SIGNAL

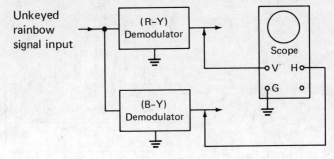

(a) Vectorgram test setup.

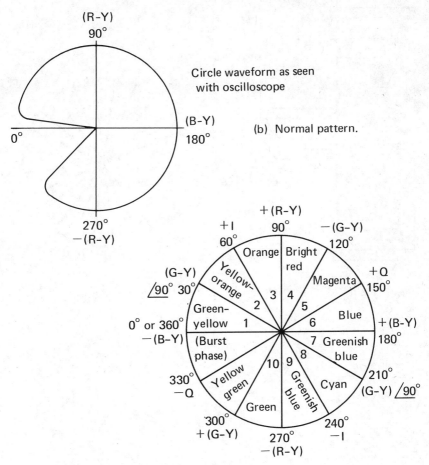

Circle waveform as seen with oscilloscope

(b) Normal pattern.

(c) Chroma phases in the vectorgram pattern.

Figure 126. Vectorgram display with unkeyed rainbow signal.

IDEAL VECTORGRAM WITH KEYED RAINBOW SIGNAL

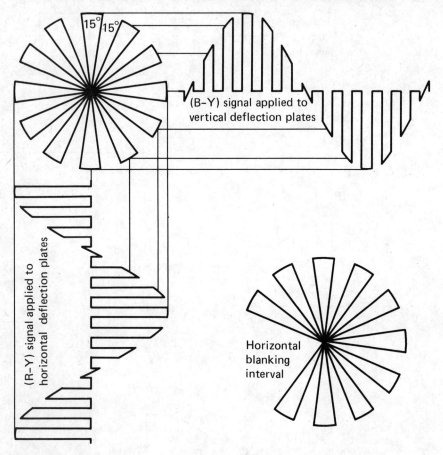

15° 15°

(B–Y) signal applied to
vertical deflection plates

(R–Y) signal applied to
horizontal deflection plates

Horizontal
blanking
interval

Figure 127. Ideal vectorgram with keyed rainbow signal.

TYPICAL VECTORGRAM DISPLAYS

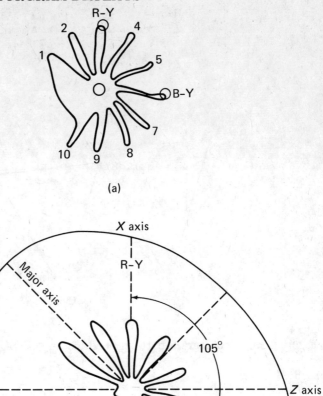

(a)

(b)

Figure 128. Typical vectorgram displays.
(a) For R–Y/B–Y receiver.
(b) For XZ receiver.

NTSC VECTORGRAM PATTERN

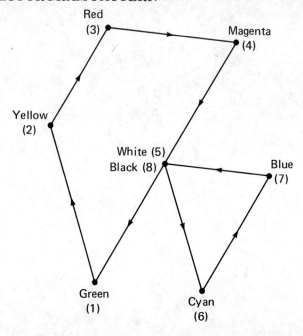

Figure 129. Vectorgram pattern produced by NTSC color-bar signal.

SECTION 11

TRANSISTOR CHARACTERISTICS

Section 11 presents transistor symbols, bipolar transistor circuit parameters, typical transistor operating voltages; input resistance versus load resistance; output resistance versus generator resistance; voltage, current, and power gain in the CB configuration; voltage, current, and power gain in the CE configuration; voltage, current, and power gain in the CC configuration; relative hybrid parameter values in the CB, CE, and CC configurations; conversion formulas for hybrid parameters, in-circuit testing of bipolar transistors, identification of bipolar transistor terminals, testing JFET transistors, and testing MOSFET transistors.

TRANSISTOR SYMBOLS

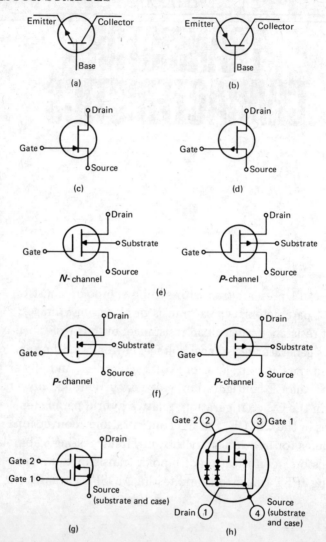

Figure 130. Transistor symbols.
   (a) Symbol for *NPN* bipolar transistor.
   (b) Symbol for *PNP* bipolar transistor.
   (c) Symbol for junction–gate, *N*–channel, field–effect transistor.
   (d) Symbol for junction–gate, *P*–channel, field–effect transistor.
   (e) Symbols for depletion–type MOSFET.
   (f) Symbols for enhancement–type MOSFET.
   (g) Symbol for dual–gate, *N*–channel, depletion–type MOSFET.
   (h) Symbol for dual–gate–protected, *N*–channel, depletion–type MOSFET.

BIPOLAR TRANSISTOR CIRCUIT PARAMETERS

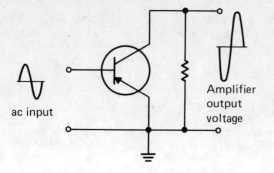

Voltage gain: 250 times
Current gain: 50 times
Power gain: 41 dB
Input resistance: 1.3KΩ
Output resistance: 6.5KΩ
(for generator internal
resistance of 1KΩ)

(a) Typical circuit values for common emitter amplifier.

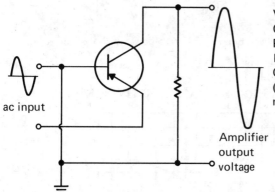

Voltage gain: 400 times
Current gain: 0.98
Power gain: 26 dB
Input resistance: 50 Ω
Output resistance: 200KΩ
(for generator internal
resistance of 1KΩ)

(b) Typical circuit values for common base amplifier.

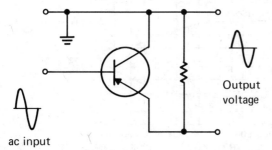

Voltage gain: 1
Current gain: 51 times
Power gain: 17 dB
Input resistance: 250KΩ
Output resistance: 500 Ω
(for generator internal
resistance of 1KΩ)

(c) Typical circuit values for common collector amplifier.

Figure 131. Bipolar transistor circuit parameters.

TYPICAL TRANSISTOR OPERATING VOLTAGES

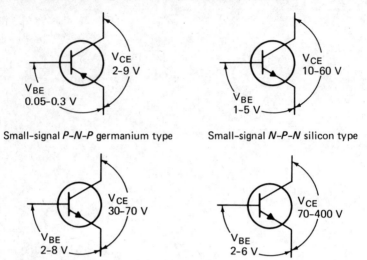

Small-signal *P-N-P* germanium type   Small-signal *N-P-N* silicon type

Medium-power *N-P-N* silicon type   High-power *N-P-N* silicon type

Figure 132. Typical transistor operating voltages.

INPUT RESISTANCE VERSUS LOAD RESISTANCE

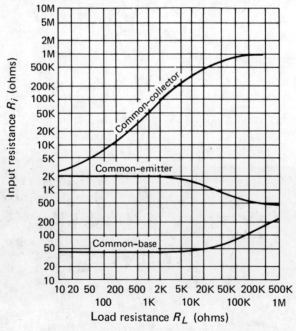

Figure 133. Input resistance versus load resistance, bipolar transistor.

OUTPUT RESISTANCE VERSUS GENERATOR RESISTANCE

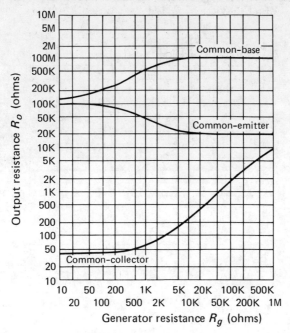

Figure 134. Output resistance versus generator resistance, bipolar transistor.

VOLTAGE, CURRENT, AND POWER GAIN, CB CONFIGURATION

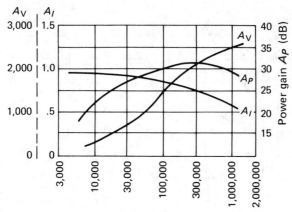

A_V = voltage gain
A_I = current gain

Figure 135. Voltage, current, and power gain, CB configuration, bipolar transistor.

VOLTAGE, CURRENT, AND POWER GAIN, CE CONFIGURATION

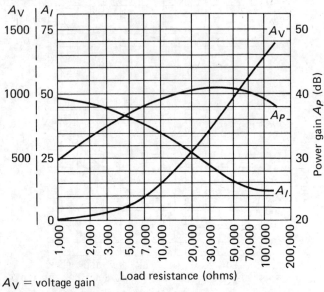

A_V = voltage gain
A_I = current gain

Figure 136. Voltage, current, and power gain, CE configuration, bipolar transistor.

VOLTAGE, CURRENT, AND POWER GAIN, CC CONFIGURATION

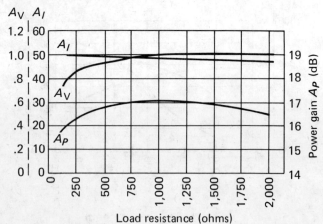

A_V = voltage gain
A_I = current gain

Figure 137. Voltage, current, and power gain, CC configuration, bipolar transistor.

108

RELATIVE HYBRID PARAMETER VALUES, CB, CE, AND CC CONFIGURATIONS

Common Emitter	Common Base	Common Collector
$h_{se} = 1{,}950\Omega$	$h_{sb} = 39\Omega$	$h_{ic} = 1950\Omega$
$\mu_{re} = 575 \times 10^{-6}$	$\mu_{rb} = 380 \times 10^{-6}$	$\mu_{rc} = 1$
$\alpha_{fe} = 49$	$\alpha_{fb} = -0.98$	$\alpha_{fc} = -50$
$h_{oe} = 24.5\,\mu\,\text{mhos}^{*}$	$h_{ob} = 0.49\,\mu\,\text{mhos}^{*}$	$h_{oc} = 24.5\,\mu\,\text{mhos}^{*}$

*or Seimen

Figure 138 Relative hybrid parameter values, CB, CE, and CC configurations.

CONVERSION FORMULAS FOR HYBRID PARAMETERS

From CE to CB	From CE to CC	From CB to CE	From CB to CC
$h_{ib} = \dfrac{h_{ie}}{1+\alpha_{fe}}$	$h_{ic} = h_{ie}$	$h_{ie} = \dfrac{h_{ib}}{1+\alpha_{fb}}$	$h_{ic} = \dfrac{h_{ib}}{1+\alpha_{fb}}$
$\mu_{rb} = \dfrac{h_{ie}h_{oe}}{1+\alpha_{fe}} - \mu_{re}$	$\mu_{rc} = 1 - \mu_{re} \cong 1$	$\mu_{re} = \dfrac{h_{ib}h_{ob}}{1+\alpha_{fe}} - \mu_{rb}$	$\mu_{rc} = \dfrac{1 = h_{ib}h_{ob}}{1+\alpha_{fb}} + \mu_{rb}$
$\alpha_{fb} = \dfrac{-\alpha_{fe}}{1+\alpha_{fe}}$	$\alpha_{fc} = -(1+\alpha_{fe})$	$\alpha_{fc} = \dfrac{-\alpha_{fb}}{1+\alpha_{fb}}$	$\alpha_{fc} = \dfrac{-1}{1+\alpha_{fb}}$
$h_{ob} = \dfrac{h_{oe}}{1+\alpha_{fe}}$	$h_{oc} = h_{oe}$	$h_{oe} = \dfrac{h_{ob}}{1+\alpha_{fb}}$	$h_{oc} = \dfrac{h_{ob}}{1+\alpha_{fb}}$

Figure 139. Conversion formulas for hybrid parameters.

IN–CIRCUIT TESTING OF BIPOLAR TRANSISTORS

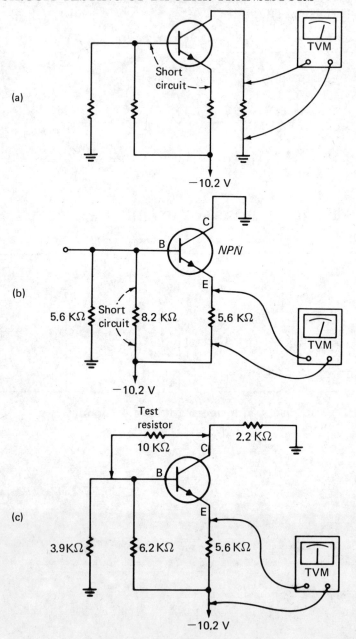

Figure 140. In-circuit testing of bipolar transistors.
(a) Turn–off test; TVM reads zero if transistor is normal.
(b) Turn–off test; TVM reads zero if transistor is normal.
(c) Turn–on test; TVM reading increases if transistor is normal.

IDENTIFICATION OF BIPOLAR TRANSISTOR TERMINALS

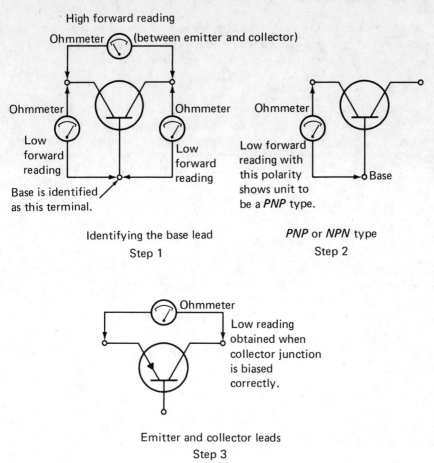

Identifying the base lead
Step 1

PNP or NPN type
Step 2

Emitter and collector leads
Step 3

Figure 141. Identification of bipolar transistor terminals.

TESTING JFET TRANSISTORS

The forward resistance of a JFET transistor can be checked with a low–voltage ohmmeter, preferably on the R X 100 range. The negative lead of the ohmmeter is connected to the gate, and the positive lead is connected to the drain or source if it is an N–channel JFET. If it is a P–channel JFET, reverse the test leads. Next, to check the reverse resistance of an N–channel JFET, connect the positive lead of the ohmmeter to the gate and connect the negative lead to the drain or source. A normal JFET reads almost infinite resistance.

TESTING MOSFET TRANSISTORS

Do not attempt to remove from the circuit or handle an FET unless it is a JFET or a gate-protected MOSFET. When handling an unprotected MOSFET, all leads should be kept shorted together by metal springs or by conductive foam. The operator's hand should be at ground potential. Tips of soldering irons should be grounded. The forward resistance and reverse resistance of a MOSFET can be checked with a low-voltage ohmmeter on its highest resistance range. The resistance reading from gate to drain or from gate to source should be practically infinite on both forward and reverse resistance tests.

SECTION 12
COUPLED CIRCUITS

Section 12 presents relations among the mutual inductance of a pair of coils and their coupling coefficient and the bandwidth of the associated tuned transformer, plus considerations of critical coupling and optimum Q value. Procedure is also outlined for the measurement of mutual inductance with an inductance bridge.

COUPLED CIRCUITS

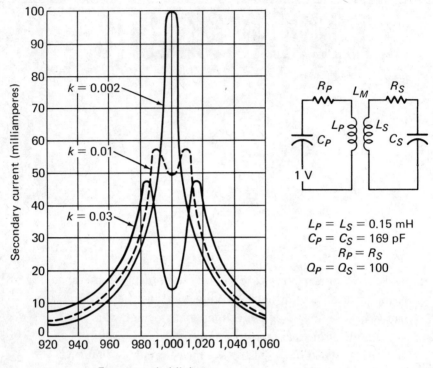

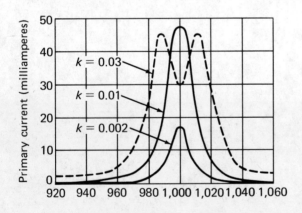

Figure 142. Primary and secondary characteristics of a typical tuned RF transformer, for three coefficients of coupling.

Inductively coupled tuned circuits with a coefficient of coupling on the order of 1 percent are often used. When two coils are inductively coupled to provide transformer action, the coupling coefficient k is given by the equation

$$k = \frac{L_m}{\sqrt{L_1 L_2}}$$

where
$$k = \text{coupling coefficient}$$
$$L_m = \text{mutual inductance}$$
$$L_1 \text{ and } L_2 = \text{inductances of the two coils}$$

Although the mathematical analysis of tuned transformer action is comparatively involved, the graphical presentation shown in Figure 142 provides pertinent data for the majority of practical situations. Typical RF transformers employ primary and secondary coils with a Q value of approximately 100. A k value of 1 percent provides a secondary bandwidth of 25 kHz. If a greater bandwidth is required in a particular application, a resistor of suitable value may be shunted across the primary, as well as across the secondary. Resistance loading reduces the output signal amplitude but has the advantage of avoiding the excessive sag in the top of the response curve that results from using high values of k. The equivalent circuit for a double-tuned transformer shown in Figure 143 is helpful in evaluating transformer response. Figure 144 is a helpful comparative chart that depicts the relative frequency responses for a single-tuned circuit versus inductively coupled circuits with four different kQ products. This chart provides response data at any operating frequency.

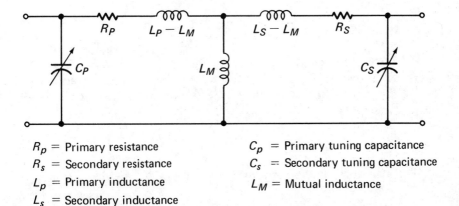

R_p = Primary resistance $\quad\quad C_p$ = Primary tuning capacitance
R_s = Secondary resistance $\quad C_s$ = Secondary tuning capacitance
L_p = Primary inductance $\quad\quad L_M$ = Mutual inductance
L_s = Secondary inductance

Figure 143. Equivalent circuit for a double-tuned transformer.

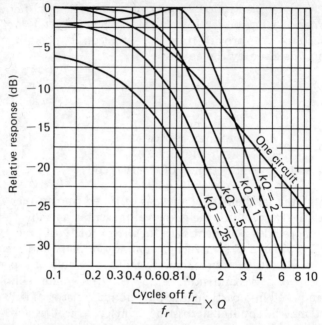

Figure 144. Relative responses for a single-tuned circuit versus typical inductively coupled circuits.

The *mutual inductance* of a transformer is generally measured with an inductance bridge, using the series–aiding and series–opposing connections of primary and secondary windings as shown in Figure 145. When the windings are connected in series–aiding, the mutual inductance adds to the intrinsic inductance value of the primary and of the secondary with the result that the total measured inductance value is

$$L\text{ total} = L_p + L_s + 2L_M$$

Next, when the windings are connected in series–opposing, the mutual inductance subtracts from the intrinsic inductance value of the primary and of the secondary with the result that the total measured inductance value is

$$L_{\text{total}} = L_p + L_s - 2L_M$$

When the foregoing equations are solved for L_M, we obtain

$$L_M = \frac{L_{\text{aiding}} - L_{\text{opposing}}}{4}$$

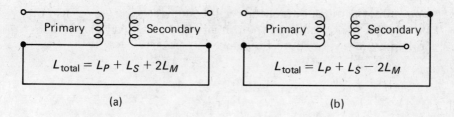

Figure 145. Primary and secondary connections for measurement of
   mutual inductance.
   (a) Series–aiding connection.
   (b) Series–opposing connection.

   The coefficient of coupling is defined as the ratio of mutual inductance to the maximum possible value of mutual inductance. This
maximum value is given by the equation

$$L_{M\,(max)} = L_p\,L_s$$

   When $L_p = L_s$, as is usually the case, the maximum possible value
of mutual inductance is equal to L_p (or L_s). The coefficient of coupling is given by the equation

$$k = \frac{L_M}{\sqrt{L_P L_S}} \quad \text{for the general case}$$

or

$$k = \frac{L_M}{L_p} = \frac{L_M}{L_s} \qquad \text{when } L_p = L_s$$

   Critical coupling provides peak output amplitude, just before the
single–humped curve changes into a double–humped curve (Figure
142). The coefficient of critical coupling is given by the equation

$$k_{(critical)} = \frac{1}{Q_p\,Q_s}$$

   Considerations of optimum Q value are depicted in Figure 146.
Most applications are concerned with reasonably flat–topped frequency response over a certain bandwidth. If the transformer windings have an excessively high Q value, the required coefficient of
coupling for the given bandwidth causes a serious sag in the center of
the frequency response curve. When an optimum Q value is employed, the output amplitude is less, but the frequency response has
an essentially flat top. On the other hand, if the Q value is less than

117

optimum, the output amplitude will be unnecessarily low. As noted previously, the Q values of the primary and secondary can be controlled by means of shunt resistance.

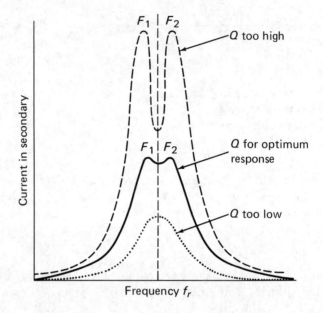

Figure 146. Effect of Q on the response characteristics of a tuned transformer.

SECTION 13

FILTER NETWORKS

Section 13 presents the design data for low–pass, high–pass, and bandpass filters of the constant–k type, with or without m–derived end sections. The general characteristics of T, π, and L sections are noted, with the effect of more than one section on the attenuation characteristic. The use of an m–derived center section is also noted.

SECTION 13

LOW-PASS FILTERS

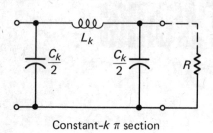

Constant-*k* π section

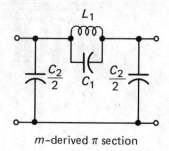

m-derived π section

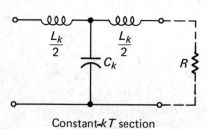

Constant-*k T* section

$$L_k = \frac{R}{\pi f_c}$$

$$C_k = \frac{1}{\pi f_c R}$$

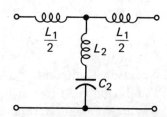

m-derived *T* section

$$L_1 = mL_k$$

$$C_1 = \frac{1-m^2}{4m}\, C_k$$

$$L_2 = \frac{1-m^2}{4m}\, L_k$$

$$C_2 = m\, C_k$$

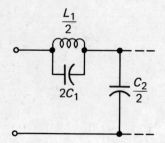

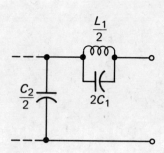

m-derived end sections for use
with intermediate π section

120

m-derived end sections for use
with intermediate T section

$$L_1 = mL_k$$

$$L_2 = \frac{1-m^2}{4m} L_k$$

$$C_1 = \frac{1-m^2}{4m} C_k$$

$$C_2 = mC_k$$

f_c = Cutoff frequency in hertz
R = Load resistance in ohms
L = Inductance in henries
C = Capacitance in farads
See text for additional details.

Figure 147. Low-pass filter sections with required L and C
values for specified cutoff frequency and load
resistance.

A low-pass filter is an LC network with a purely resistive load that
will permit all frequencies below a chosen cutoff frequency to pass
through with little or no loss, whereas all frequencies above the cutoff frequency are attenuated or rejected. The pass band is the frequency spectrum that is passed with little or no loss. The stop band
is the frequency interval over which signals are attenuated or rejected.
A resistive load for a filter is often called its terminating impedance.
Note that even a small amount of a reactance in the terminating impedance can distort the filter frequency characteristic considerably.

The basic configurations are called the T and π (pi) sections. Either
configuration will provide the same frequency characteristic. The
simplest T and π sections are called the constant-k type, and they are
in very wide use. As shown in Figure 148 (a), a T section may be fur-
ther simplified into an L section. If a single L, T, or π section is used,
the attenuation characteristic is gradual, as seen at A in Figure 148 (c).
On the other hand, if three sections are connected in series, the attenuation characteristic is abrupt, as seen at B in Figure 148 (c). In general,
the more sections that are connected in series, the more abrupt is the
attenuation characteristic. (There must be no coupling between any
of the inductors.)

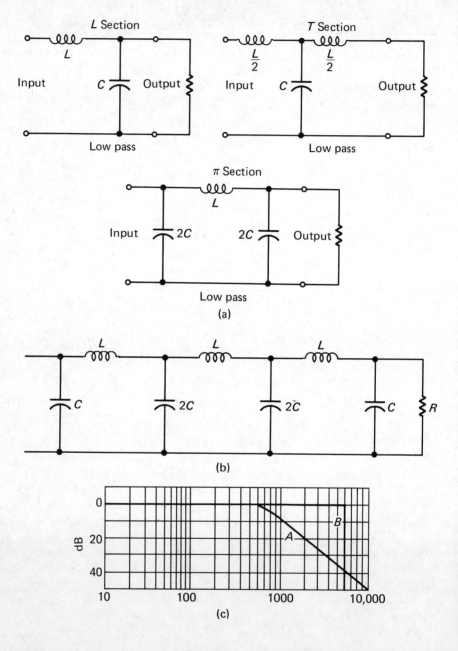

Figure 148. Constant-k low-pass filter characteristics.
(a) L, T, and π sections.
(b) Three π sections connected in series.
(c) Frequency response for one section shown at A and frequency response for three sections shown at B.

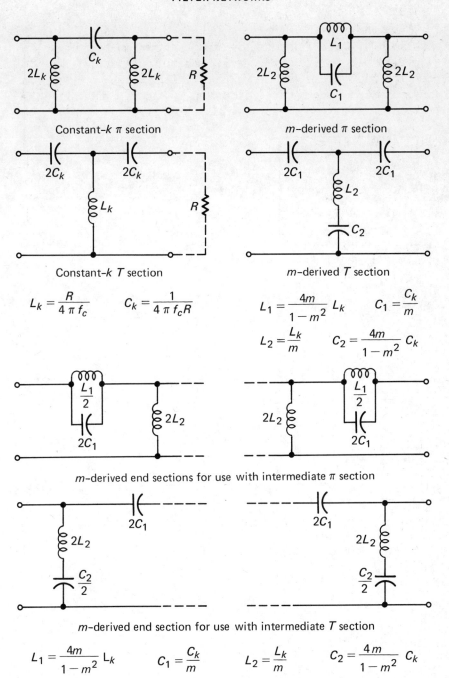

Constant-k π section

m-derived π section

Constant-k T section

m-derived T section

$$L_k = \frac{R}{4\pi f_c} \qquad C_k = \frac{1}{4\pi f_c R}$$

$$L_1 = \frac{4m}{1-m^2} L_k \qquad C_1 = \frac{C_k}{m}$$

$$L_2 = \frac{L_k}{m} \qquad C_2 = \frac{4m}{1-m^2} C_k$$

m-derived end sections for use with intermediate π section

m-derived end section for use with intermediate T section

$$L_1 = \frac{4m}{1-m^2} L_k \qquad C_1 = \frac{C_k}{m} \qquad L_2 = \frac{L_k}{m} \qquad C_2 = \frac{4m}{1-m^2} C_k$$

Figure 149. High–pass filter sections with required L and C values for specified cutoff frequency and load resistance.

When several sections are connected in series, optimum cutoff characteristics can be obtained by using m-derived end sections before and after a constant-k center section, although an m-derived center section can also be used. Figure 149 shows the configurations, with required L and C values. Note that the factor m determines the

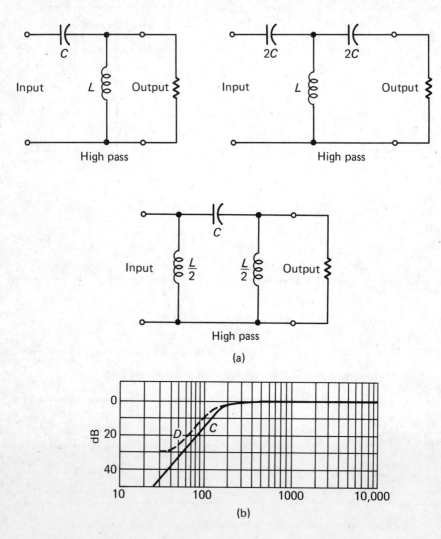

Figure 150. Constant-k high-pass filter characteristics.

(a) L, T, and π sections.

(b) Frequency response for one section shown at C, using high-Q inductor(s); effect of reduced Q value shown at D.

ratio of the cutoff frequency f_c to a frequency of high attenuation. When only one m-derived section is to be used, a value of 0.6 is generally chosen for m. Regardless of the filter configuration, f_c is always defined as the frequency at which the response is down to 70.7 percent of maximum.

HIGH-PASS FILTERS

A high-pass filter is an LC network with a purely resistive load that will permit all frequencies above a chosen cutoff frequency to pass through with little or no loss, whereas all frequencies below the cutoff frequency are attenuated or rejected. Figure 149 shows basic high-pass filter sections, with required L and C values for specified cutoff frequency and load-resistance values. Figure 150 shows the characteristics for single high-pass filter sections, using high-Q and medium-Q inductors. As explained above, a more abrupt cutoff characteristic can be obtained by connecting several high-pass sections in series. A still more abrupt cutoff characteristic can be obtained by utilizing m-derived end sections. (There must be no coupling between any of the inductors.)

BANDPASS FILTERS

A bandpass filter is an LC network with a purely resistive load that will permit frequencies between two selected cutoff frequencies to pass through with little or no loss, whereas all frequencies outside of this pass band are attenuated or rejected. Figure 151 shows basic bandpass filter sections, with required L and C values for specified cutoff frequencies and load-resistance values. Figure 152 depicts typical response curves for bandpass filters. Curve E illustrates the response of a low-Q, narrow-band filter section, curve F shows the response for a high-Q, narrow-band filter section, and G shows the typical response of a wide-band, high-Q, two-section filter. More abrupt cutoff characteristics can be obtained by utilizing additional sections connected in series.

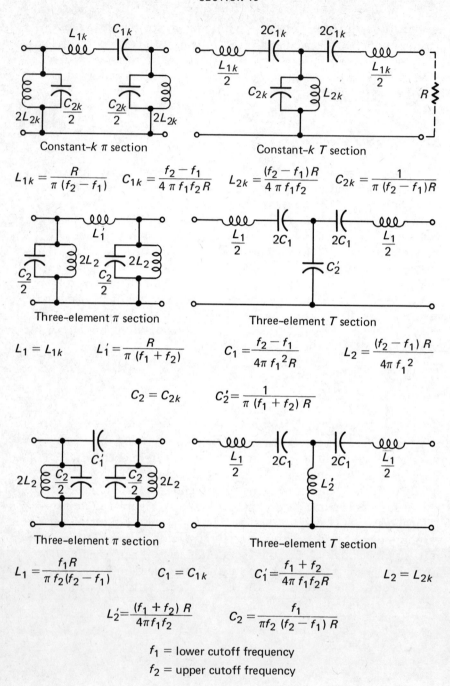

$$L_{1k} = \frac{R}{\pi \, (f_2 - f_1)} \qquad C_{1k} = \frac{f_2 - f_1}{4 \, \pi \, f_1 f_2 R} \qquad L_{2k} = \frac{(f_2 - f_1) R}{4 \, \pi \, f_1 f_2} \qquad C_{2k} = \frac{1}{\pi \, (f_2 - f_1) R}$$

$$L_1 = L_{1k} \qquad L_1' = \frac{R}{\pi \, (f_1 + f_2)} \qquad C_1 = \frac{f_2 - f_1}{4\pi \, f_1{}^2 R} \qquad L_2 = \frac{(f_2 - f_1) \, R}{4\pi \, f_1{}^2}$$

$$C_2 = C_{2k} \qquad C_2' = \frac{1}{\pi \, (f_1 + f_2) \, R}$$

$$L_1 = \frac{f_1 R}{\pi \, f_2 (f_2 - f_1)} \qquad C_1 = C_{1k} \qquad C_1' = \frac{f_1 + f_2}{4\pi \, f_1 f_2 R} \qquad L_2 = L_{2k}$$

$$L_2' = \frac{(f_1 + f_2) \, R}{4\pi f_1 f_2} \qquad C_2 = \frac{f_1}{\pi f_2 \, (f_2 - f_1) \, R}$$

$$f_1 = \text{lower cutoff frequency}$$
$$f_2 = \text{upper cutoff frequency}$$

Figure 151. Basic bandpass filter section configurations, with required L and C values.

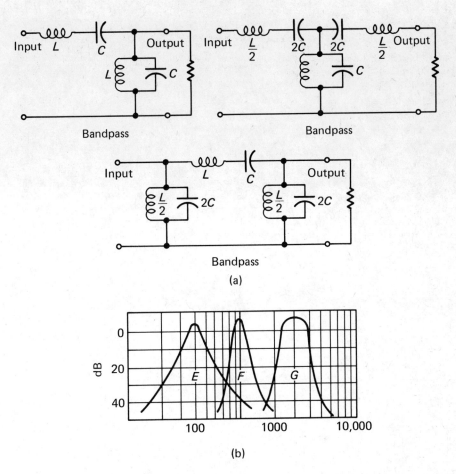

Figure 152. Constant-k bandpass filter characteristics.
   (a) Basic sectional configurations.
   (b) Frequency responses; E, low-Q narrow-band filter section;
       F, high-Q narrow-band filter section; G, wide-band, high-Q,
       two-section filter.

127

SECTION 14

LC MATCHING NETWORKS

Section 14 presents the basic characteristics of *LC* matching networks, with practical design data. The relation of the *Q* value to the reactance–resistance ratio is explained, with the condition for maximum power transfer. Efficiency considerations are noted, plus the application of tapped tuned circuits.

Impedance-matching networks are used to obtain maximum power transfer, as from a radio transmitter to an antenna transmission line. Basic impedance-matching networks with required parameters are shown in Figure 153. Note that the Q values in the matching networks must be selected in accordance with the equa-

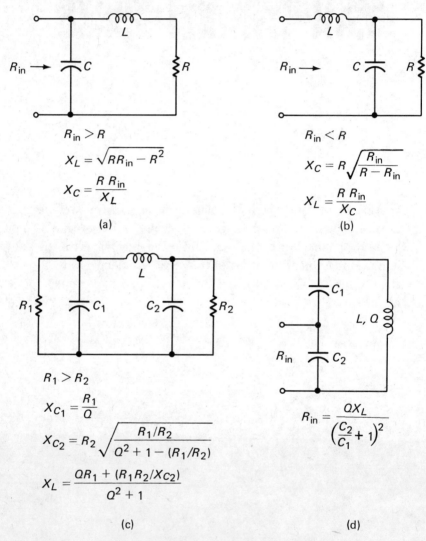

$$R_{in} > R$$
$$X_L = \sqrt{RR_{in} - R^2}$$
$$X_C = \frac{R\,R_{in}}{X_L}$$

(a)

$$R_{in} < R$$
$$X_C = R\sqrt{\frac{R_{in}}{R - R_{in}}}$$
$$X_L = \frac{R\,R_{in}}{X_C}$$

(b)

$$R_1 > R_2$$
$$X_{C_1} = \frac{R_1}{Q}$$
$$X_{C_2} = R_2\sqrt{\frac{R_1/R_2}{Q^2 + 1 - (R_1/R_2)}}$$
$$X_L = \frac{QR_1 + (R_1 R_2/X_{C_2})}{Q^2 + 1}$$

(c)

$$R_{in} = \frac{QX_L}{\left(\dfrac{C_2}{C_1} + 1\right)^2}$$

(d)

Figure 153. Basic impedance-matching networks.
            (a) L network, high Z to low Z.
            (b) L network, low Z to high Z.
            (c) π network, high Z to low Z.
            (d) Tapped tuned circuit, R_{in} less than circuit Z.

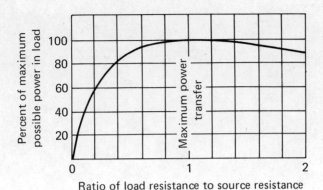

Figure 154. Power transfer versus ratio of load resistance to source resistance.

tions given. In Figure 153 (a), the Q value is equal to X_L/R, or to R_{in}/X_C. In other words, when values of R_{in} and R are assigned, particular values for L and C follow from the calculated values of X_L and X_C at the operating frequency. In Figure 153 (b), the Q value is equal to X_L/R_{in}, or to R/X_C. In Figure 153 (c), the Q value is equal to R_1/X_{C1}. Tapped tuned circuits for impedance transformation are used principally in receivers.

Maximum power transfer is obtained from a source to a load when the effective ratio of load resistance to source resistance is equal to 1. The effect of mismatching is shown in Figure 154. At maximum power transfer, the system efficiency is 50 percent. In most communication systems, maximum power transfer rather than high efficiency is desired. Figure 155 shows the relation of efficiency to power transfer. Maximum efficiency occurs when minimum power is transferred from source to load.

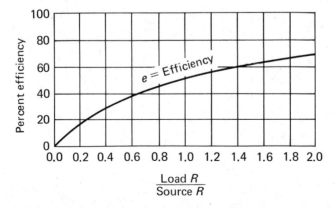

Figure 155. Percent efficiency versus ratio of load resistance to source resistance.

SECTION 15

NOISE MEASUREMENTS

Section 15 presents a general discussion of noise and its measurement procedures. Weighting is explained, and loudness units are defined. The application of a VU meter is noted. A diagram of the relation between loudness units and phons is provided in order to show the recognized objective-subjective relations that have been established.

Noise is unwanted sound. It may be measured by a sound level meter or electrically. The indicating meter may have a flat frequency response, or it may be used with a weighting network. If a weighting network is utilized, there will be substantial attenuation of the lower frequencies, including any hum that might be present. Therefore the frequency response of the meter should always be stated in combination with a noise measurement. Typical weighting networks follow the 40-dB equal-loudness contour or the 70-dB equal-loudness contour of the Fletcher and Munson curves depicted in Figure 155. Note that the phon is the unit of loudness level. The loudness level in phons is equal to the dB level at a frequency of 1 kHz only.

The measurement of noise in amplifiers for sound equipment denotes a measurement of steady-state noise and of pulse noise waves. The noise level is defined as the level of any noise signals appearing at the output terminals with no signal applied to the input. In general, the weighted noise level is in accordance with the 70-dB equal-loudness contour (Figure 156) and is expressed in dBm units. A standard VU (volume-units) meter is utilized. A properly weighted amplifier for the meter may be constructed by employing an *RC* differentiating network having a 1 ms time constant with an amplifier having

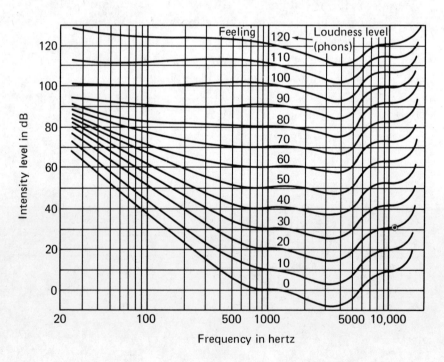

Figure 156. Fletcher and Munson contours of equal loudness.

a frequency response of ± 1 dB from 50 to 15,000 Hz. This network provides an attenuation of 1 dB at 300 Hz, 5.7 dB at 100 Hz, and 9 dB at 60 Hz.

For the purpose of noise measurements, a *loudness unit* (LU) has been established. The loudness unit is based on the need for a unit that is proportional to the sensation of loudness. In other words, if the number of loudness units is doubled, a sensation of twice the original loudness is provided. Figure 157 shows the relation between loudness units and phons.

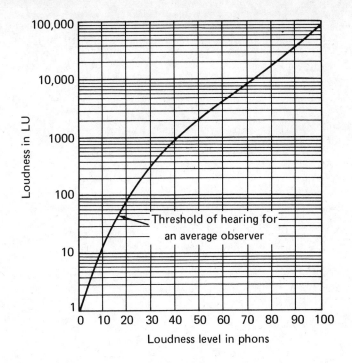

Figure 157. Relation between loudness units and phons.

SECTION 16
NETWORK THEOREMS

Section 16 presents the basic network theorems comprising
Thévenin's theorem, Norton's theorem, the Superposition
theorem, and the Reciprocity theorem, with an explanation
of the derivation of an equivalent T section and the derivation
of an equivalent π section.

THÉVENIN'S THEOREM

Thévenin's theorem states that any linear network of resistances and sources, if viewed from any two points in the network, can be replaced by an equivalent resistance R_{Th} in series with an equivalent voltage source E_{Th}. Thus any linear dc circuit, regardless of complexity, can be replaced by a Thévenin equivalent circuit, as depicted in Figure 158. For example, assume that the circuit in Figure 159 is to be analyzed by means of Thévenin's theorem. We proceed to evaluate E_{Th} and R_{Th} as follows:

1. Disconnect the section of the circuit considered to be the load—R_L in Figure 159 (a).

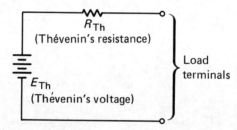

Figure 158. A Thévenin equivalent circuit.

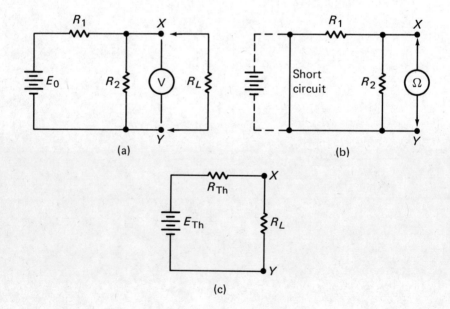

Figure 159. Deriving a Thévenin equivalent circuit.

138

2. By measurement or calculation, determine the voltage that would appear between the load terminals with the load disconnected (terminals X and Y). This open–circuit voltage is called the Thévenin voltage, E_{Th}.
3. Replace each voltage source within the circuit by its internal resistance. A constant–voltage source is replaced by a short circuit. This step is depicted in Figure 159 (b).
4. By measurement or calculation, determine the resistance the load would "see," looking back into the network from its load terminals; this is the Thévenin resistance, R_{Th} in Figure 159 (b).
5. Draw the equivalent circuit consisting of R_L and R_{Th} in series, connected across E_{Th}, as shown in Figure 159 (c). Solve for the load current and voltage.

In a similar manner, Thévenin's theorem applies to any ac network composed of linear impedances.

NORTON'S THEOREM

Norton's theorem is a restatement of Thévenin's theorem for a constant–current source. It states that any linear network of resistances and sources, if viewed from any two terminals in the network, can be replaced by an equivalent resistance R_N in shunt with an equivalent constant–current source I_N. Figure 160 shows a Norton equivalent circuit. Note that R_N has the same value as R_{Th} in a Thévenin analysis. The procedure for simplification of a circuit to a Norton equivalent circuit is exemplified in Figure 161, page 140.

1. Disconnect R_L, as shown in Figure 161 (a).
2. By measurement or calculation, determine the current that would flow through a short circuit between A and B, as shown in Figure 161 (b). This is I_N.
3. Remove the physical or imaginary short circuit across the load.
4. Replace each source within the network with its internal resistance (this is zero ohms for a constant–voltage source). The circuit shown in Figure 161 (c) is thereby obtained.

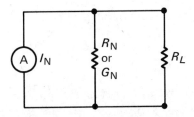

Figure 160. A Norton equivalent circuit.

4. By measurement or calculation, determine the resistance R_N "looking back"; this is the same as R_{Th}. If desired, we may state the reciprocal of this value as the conductance G_N "looking back" into the network.
5. Draw the equivalent circuit comprising R_L, R_N, and I_N all in parallel, as shown in Figure 161 (a), and solve for the load conditions.

If we had started with a constant-current source in the network, we would measure I_N as before but would replace the source with a current generator that had infinite internal resistance (zero conductance). Again, if we started with an energy source that was neither a constant-voltage source nor a constant-current source, we would measure I_N as before, but we would replace the source with a resistance equal to the internal resistance of the source.

In a similar manner, Norton's theorem applies to any ac circuit consisting of linear impedances.

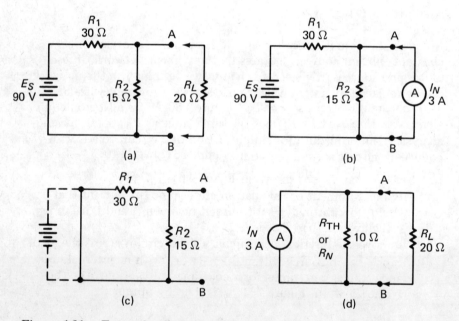

Figure 161. Example of the application of Norton's theorem.

SUPERPOSITION THEOREM

The Superposition theorem states that, in any linear network consisting of linear resistances, the current at any point is equal to the algebraic sum of the currents that would be present if each source

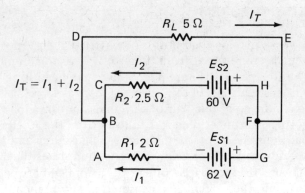

Figure 162. A configuration that consists of two sources supplying a common load.

were considered separately, all other sources being replaced at this time by resistances equal to their internal resistances. For example, with reference to Figure 162, a network of three resistors and two constant-voltage sources will be analyzed by means of the Super-position theorem. Note that in this example the source polarities are such that I_T is equal to the sum of I_1 and I_2.

We proceed by drawing the equivalent circuit shown in Figure 163 (a), with the voltage source E_{S2} replaced by its internal resistance, and solve for I_A, which is equal to 5.63 A. Next, we draw the equivalent circuit shown in Figure 163 (b), with the voltage source E_{S1} replaced by its internal resistance, and solve for I_B, which is equal to 4.38 A, approximately. In turn, the load current is equal to the sum of I_A and I_B, or 10 A, approximately.

In a similar manner, the Superposition theorem applies to any ac circuit comprising linear impedances.

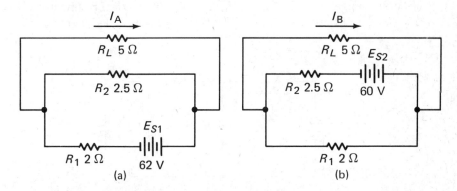

Figure 163. Equivalent circuits for application of the Superposition theorem.

RECIPROCITY THEOREM

Various networks may be used in such a way that their input and output terminals are reversed from time to time, as in communication systems. The Reciprocity theorem eliminates the need for a pair of solutions in this situation. In other words, if the network is solved for a source condition at one pair of terminals and for a load condition at the other pair of terminals, it follows from the Reciprocity theorem that the same solution holds true if the source and load terminals are reversed. The Reciprocity theorem states that, in any system of linear resistances, if a voltage E_S is applied between any two terminals, and the current I is measured in any branch, their quotient will be equal to the quotient obtained when the positions of E and I are reversed.

For example, with reference to Figure 164, a reciprocity analysis is made as follows. If we form the quotient E_S/I_2, it is demonstrable that this quotient remains the same when the battery and ammeter are interchanged. Thus we write

$$I_1 = \frac{E_{S3}}{R_1 + R_2 R_3 / R_2 + R_3}$$

$$I_2 = I_1 \frac{R_p}{R_2} = I_1 \left(\frac{R_3}{R_2 + R_3} \right)$$

$$\frac{E_S}{I_2} = \frac{R_1 R_2 + R_1 R_3 + R_2 R_3}{R_3}$$

Next, if we interchange the battery and ammeter in Figure 164, this circuit change corresponds to interchanging R_1 and R_2 in the foregoing equations. Since interchanging R_1 and R_2 in the equations does not change the value of E_S/I_2, the demonstration has been made for the exemplified circuit. The quotient E_S/I_2 is called the transfer

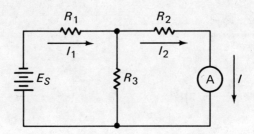

Figure 164.   A series–parallel circuit for reciprocity analysis demonstration.

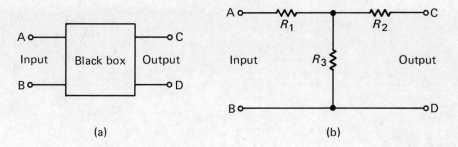

Figure 165. Black box and equivalent T section for any linear bilateral network.

resistance of the circuit and is numerically equal to the quotient of input voltage and output current. Finally, it can be demonstrated that the Reciprocity theorem holds true for any linear bilateral network because its truth has been demonstrated for a T section in Figure 164, and any linear bilateral network can be reduced to an equivalent T section.

Similarly, the Reciprocity theorem applies to any ac circuit comprised of linear impedances.

EQUIVALENT T SECTION

Any linear bilateral network can be reduced to an equivalent T section. With reference to Figure 165, we may make resistance measurements at the input and output terminals of the black box, which represents any linear bilateral network. These resistance measurements may be made with the opposite pair of terminals either open-circuited or short–circuited. In turn, to find the values of R_1, R_2, and R_3 for the equivalent T section, we write

$$R_1 = R_{ABO} - \sqrt{R_{CDO}(R_{ABO} - R_{ABS})}$$

$$R_2 = R_{CDO} - \sqrt{R_{CDO}(R_{ABO} - R_{ABS})}$$

$$R_3 = \sqrt{R_{CDO}(R_{ABO} - R_{ABS})}$$

These relations apply also to ac circuits composed of linear impedances.

Note that an equivalent T section is limited in its equivalence to input and output terminal resistances; it is not necessarily true that the resistance between terminals A and C of the black box is equal to

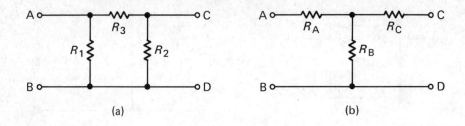

Figure 166. Equivalent circuit configurations.
(a) π section.
(b) T section.

$R_1 + R_2$, nor that the resistance between terminals B and D of the black box is equal to zero in Figure 165. That is, it is essential not to exceed the competence of a network theorem.

EQUIVALENT π SECTION

Every T section has an equivalent π section. With reference to Figure 166, R_A, R_B, and R_C can be expressed in terms of R_1, R_2, and R_3, in order to obtain another form of equivalent circuit. These expressions are as follows:

$$R_1 = \frac{R_A R_B}{R_A + R_B + R_C}$$

$$R_2 = \frac{R_B R_C}{R_A + R_B + R_C}$$

$$R_3 = \frac{R_A R_C}{R_A + R_B + R_C}$$

$$R_A = \frac{R_1 R_2 + R_2 R_3 + R_1 R_3}{R_1}$$

$$R_B = \frac{R_1 R_2 + R_2 R_3 + R_1 R_3}{R_2}$$

$$R_C = \frac{R_1 R_2 + R_2 R_3 + R_1 R_3}{R_3}$$

Figure 167 shows a memory aid for the equations of T and π sections. Note that R_1 is flanked by R_A and R_B; R_2 is flanked by R_B

144

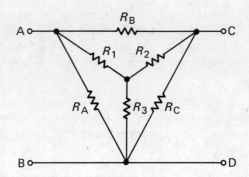

Figure 167. Memory aid for equations of T and π sections.

and R_C; R_3 is flanked by R_A and R_C. This is the same relation that is written in the numerators of the first three equations on page 144. All denominators are the same and are the sum of R_A, R_B, and R_C. Next, with reference to the second three equations, all the numerators are the same; it is the sum of the pair combinations of R_1, R_2, and R_3. Then, in order to form the denominators, we write R_1 in correspondence to R_A, R_2 in correspondence to R_B, and R_3 in correspondence to R_C.

Equivalent π sections can be calculated in a similar manner for ac circuits composed of linear impedances.

T AND H ATTENUATORS AND PADS

An attenuator is a variable section, and a pad is a fixed section. Figure 168 shows curves for basic T and H attenuators and pads; these curves often facilitate design procedures. For input–output resistance values other than those shown in the diagram, interpolation may be used for practical approximation of required values.

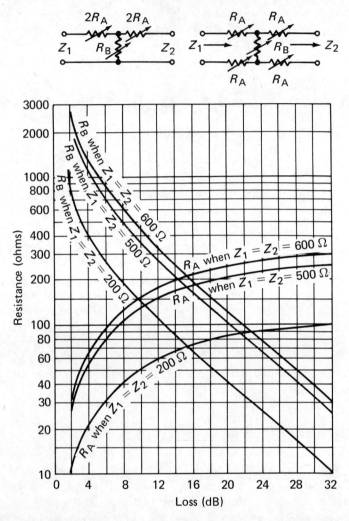

Figure 168. Resistance values for basic *T* and *H* attenuators and pads.

SECTION 17

LINE-SECTION CHARACTERISTICS

Section 17 presents the characteristics of quarter-wave and half-wave line sections. The reactances of eighth-wave sections are noted, with the general nature of reactance variation along arbitrary open-circuited and short-circuited lines. Relations of quarter-wave sections to conventional series- and parallel-resonant circuits are depicted, and impedance inversion by half-wave sections is illustrated.

Short-circuited or open-circuited line sections (Figure 169, page 149) have many of the characteristics of resonant circuits. An example of a line terminated in a short circuit is shown in the diagram. The line is a three-quarter wavelength (electrical length) in this example. Note that the voltage is low and the current is high at the load end of the line. This E/I ratio corresponds to a low resistance value. The ac voltage from the generator travels down the line and is reflected back by the short-circuit termination. In turn, there are two traveling waves on the line, which alternately reinforce and cancel each other, thus producing standing waves on the line. On the other hand, if the line were terminated in its own characteristic resistance (impedance), all the incoming power would be dissipated by the load and there would be no reflections and no standing waves.

With reference to Figure 169 (a), at a point one-half wavelength from the short-circuit termination, the voltage has shifted one-half cycle (a half wavelength), and the current has also shifted one-half cycle. The amplitude of voltage and current at this point is the same as at the short-circuited terminals. In other words, a one-half wavelength line section "repeats the load." Next, at a point one-quarter wavelength from the short-circuit termination, the voltage is maximum and the current is zero. Since the E/I ratio is infinite at this point, the input impedance here corresponds to an open circuit. In other words, a one-quarter wavelength line section "inverts the load." Again, a one-eighth length line section with an open-circuit termination "looks like" a capacitor at its input terminals. In other words, the phase of the current is shifted $45°$ leading by the line section. The capacitance value is numerically equal to the value of the characteristic impedance of the line section.

Frequency and wavelength are related by the equation

$$\lambda = \frac{c}{f}$$

where λ = wavelength in meters
c = velocity of light (300,000,000 meters per second)
f = frequency in hertz

If a one-eighth wavelength line is terminated by a short circuit, its input terminals "look like" an inductor. This inductance value is numerically equal to the value of the characteristic impedance of the line section. A quarter-wave line section can be considered to be built up from an open one-eighth wavelength section and a short-circuited one-eighth wavelength section. The input end of the quarter-wavelength section "looks like" an open circuit, or a high-Q parallel-resonant circuit. On the other hand, a quarter-wavelength section with an open-circuit termination "looks like" a short-circuit at its input end; it is equivalent to a high-Q series-resonant circuit.

148

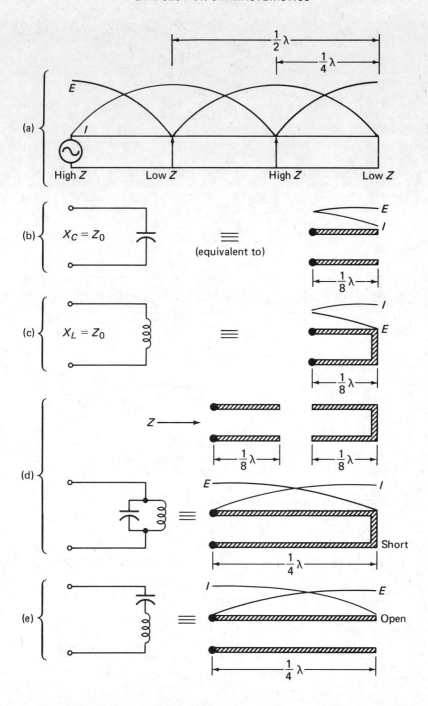

Figure 169. Short–circuited and open–circuited line sections are comparable to resonant circuits.

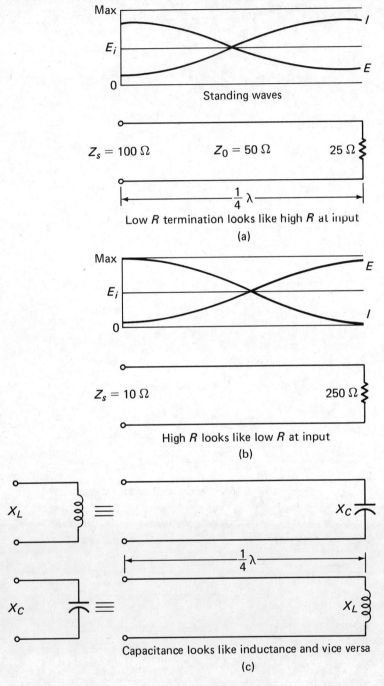

Figure 170. Impedance inversion by quarter-wavelength line-section action.

With reference to Figure 170, a quarter–wavelength line section terminated in a resistance greater than Z_0 (characteristic impedance of the line section) will have an input resistance less than Z_0. This inversion action of a quarter–wavelength line section is given by

$$\frac{Z_R}{Z_0} = \frac{Z_0}{Z_s}$$

where Z_R = terminating resistance
$\quad Z_s$ = input or sending-end impedance
$\quad Z_0$ = characteristic impedance of the line section

Whenever the termination is other than Z_0 in value, there are reflections on the line section. The value of reflected voltage is given by

$$E_R = E_i \frac{R_L - Z_0}{R_L + Z_0}$$

where E_R = reflected voltage
$\quad E_i$ = input or applied voltage
$\quad R_L$ = terminating resistance
$\quad Z_0$ = characteristic impedance of the line section

It is evident that a quarter–wavelength line section can be used to match an input impedance to an output impedance by proper choice of Z_0. This required value of Z_0 is given by

$$Z_0 = \sqrt{Z_s Z_R}$$

where Z_0 = characteristic impedance of the line section
$\quad Z_s$ = input impedance of the section
$\quad Z_R$ = output impedance of the section

SECTION 18
CIRCUIT SYMBOLS

ADJUSTABLE (CONTINUOUSLY ADJUSTABLE)

The shaft of the arrow is drawn about 45° across the body of the symbol.

AMPLIFIER

The triangle is pointed in the direction of transmission, and the symbol represents any method of amplification.

Amplifier with two inputs

Amplifier with two outputs

Amplifier with adjustable gain

ANTENNA

Dipole

Loop

ATTENUATOR (ATTENUATOR, VARIABLE)

General

Balanced, general

Unbalanced, general

AUDIBLE SIGNALING DEVICE

Bell, electrical; ringer, telephone

Buzzer

Horn, electrical; loudspeaker; siren; underwater sound projector or transceiver: General

BATTERY

The long line is always positive, but polarity may be indicated in addition.

Generalized direct-current source:

One cell

Multicell

CAPACITOR

If it is necessary to identify the capacitor electrodes, the curved element shall represent the outside electrode in fixed paper-dielectric and ceramic-dielectric capacitors, the moving element in adjustable and variable capacitors, and the low-potential element in feed-through capacitors.

Polarized capacitor

Adjustable or variable capacitor

Adjustable or variable capacitors with mechanical linkage of units

Feed-through capacitor between two inductors with third lead connected to chassis

CELL, PHOTOSENSITIVE

Indicates that the characteristic of the element within the circle varies under the influence of light.

Photoconductive resistor

Photovoltaic transducer

CIRCUIT BREAKER

CIRCUIT RETURN

Ground:
A conducting connection to a structure that serves a function similar to that of an earth ground.

Chassis or frame connection:
A conducting connection to a chassis or frame of a unit

Common connections:
Conducting connections made to one another. All like–designated points are connected.

COIL RELAY

CONNECTOR DISCONNECTING DEVICE

Female contact

Male contact

Coaxial with the outside conductor shown carried through

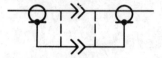

Communication switchboard-type connector:

Two-conductor (jack)

Two-conductor (plug)

Connectors commonly used for power supply purposes:

Female contact

Male contact

Two-conductor nonpolarized connector with female contacts

Two-conductor nonpolarized connector with male contacts

CONTACT, ELECTRICAL

Fixed contact for jack, key, relay, etc.

Fixed contact for switch

o or ⟶

Fixed contact for momentary switch

Sleeve

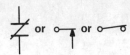

Closed contact (break)

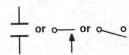

Open contact

Make–before–break

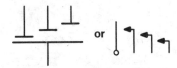

Time–sequential closing

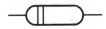

DELAY FUNCTION, DELAY LINE

General

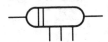

Tapped delay function

Tapped delay function

FUSE

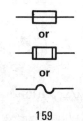

GROUND

See CIRCUIT RETURN

HALL GENERATOR

HANDSET, OPERATOR'S SET

General

With push-to-talk switch

INDUCTOR WINDING REACTOR

Either symbol may be used

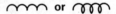

Magnetic-core inductors

Tapped

Adjustable inductor

Application; adjustable or continuously adjustable inductor

KEY, TELEGRAPH

LAMP

Ballast lamp: ballast tube

Neon:
Direct–current type

Lamp, incandescent

LOGIC

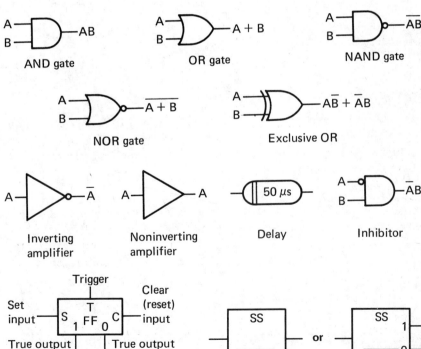

161

or

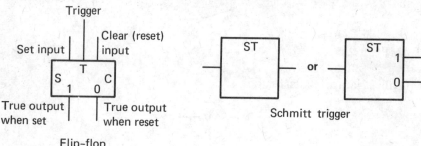

Flip-flop

Schmitt trigger

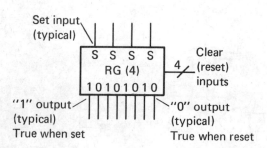

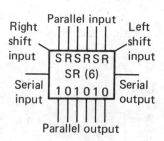

or

or

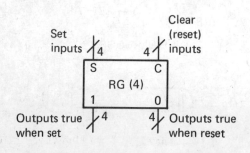

Register

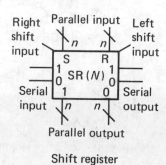

Shift register

METER

MICROPHONE

NETWORK

NET

OSCILLATOR, GENERALIZED ALTERNATING-CURRENT SOURCE

PATH, TRANSMISSION CONDUCTOR CABLE WIRING

A single line represents the entire group of conductors.

Conductive path or conductor; wire

Two conductors or conductive paths of wires

Crossing of paths or conductors not connected

Junction of connected paths

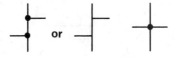

Shielded single conductor

Coaxial cable

Two-conductor cable

Shielded two-conductor cable with shield grounded

PHASE SHIFTER

General

Three-wire or three-phase

Application: adjustable

PICKUP HEAD

General

Stereo

PIEZOELECTRIC CRYSTAL UNIT

RECEIVER, TELEPHONE EARPHONE

General

Headset, double

RECTIFIER

General

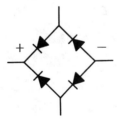

Full–wave bridge–type rectifier

REPEATER

One–way repeater:
Triangle points in the direction of transmission.

Two–wire, two–way repeater

RESISTOR

General

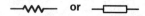

Tapped resistor

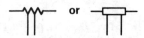

Application: with adjustable contact

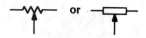

Application: adjustable or continuously adjustable (variable) resistor

SEMICONDUCTOR DEVICE

Semiconductor diode

Capacitive diode (Varactor)

Breakdown diode, unidirectional (also Backward diode)

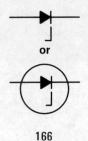

Breakdown diode, bidirectional

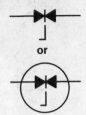

Tunnel diode

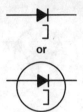

Temperature–dependent diode

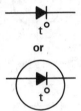

Photodiode

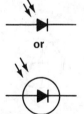

Semiconductor diode, *PNPN*-type switch

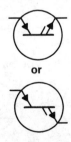

The elements of the symbol must be drawn so as to show clearly the operating function of the device.

PNP transistor

PNP transistor with one electrode connected to envelope (in this case, the collector electrode)

NPN transistor

Unijunction transistor with *N*–type base

Unijunction transistor with *P*–type base

Field–effect transistor with *N*–type base

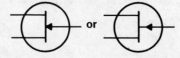

Field–effect transistor with *P*–type base

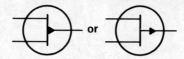

Semiconductor triode, *PNPN*-type switch

Semiconductor triode, *NPNP*-type switch

SHIELD

‒ ‒ ‒ ‒ ‒

SHIELDING (short dashes)

‒ ‒ ‒ ‒ ‒
‒ ‒ ‒ ‒ ‒

Normally used for electric or magnetic shielding. When used for other shielding, a note should so indicate.

SWITCH

Single throw, general

Double throw, general

Circuit closing (make)

Circuit opening (break)

Two-circuit

Nonlocking; momentary or spring return:
Circuit closing (make)

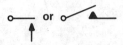

Circuit opening (break)

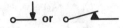

Two-circuit

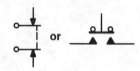

Selector or multiposition switch

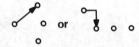

Break-before-make, nonshorting

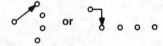

Make-before-break, shorting

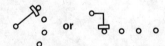

THERMAL ELEMENT

THERMOCOUPLE

TRANSFORMER

One winding with adjustable inductance

Each winding with separately adjustable inductance

Magnetic–core transformer

Shielded transformer with magnetic core shown

SECTION 19
DIGITAL LOGIC TRUTH TABLES

A truth table is a chart of a logic function listing all possible combinations of input values and indicating, for each combination, the true output values. The following truth tables are set forth in this section:

Two-input AND gate, Figure 171.

Three-input AND gate, Figure 172.

Two-input OR gate, Figure 173.

Three-input OR gate, Figure 174.

Inverter, Figure 175.

RS flip-flop, Figure 176.

RS flip-flop (NOR gate configuration), Figure 177.

RS flip-flop (NAND gates using equivalent NOR symbols), Figure 178.

Master-slave flip-flop using gated RS flip-flops, Figure 179.

Type D flip-flop, Figure 180.

Gated RS flip-flop, Figure 181.

Gated RS flip-flop (NAND gates), Figure 182.

J-K flip-flop (RST flip-flop/gates), Figure 183.

DeMorgan's equivalent basic gates, Figure 184.

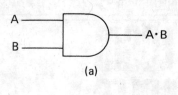

Condition	A	B	A·B
1	F	F	F
2	F	T	F
3	T	F	F
4	T	T	T

(b)

T = True

F = False

A·B = A AND B

Figure 171. Two-Input AND gate
(a) Logic symbol
(b) Truth table

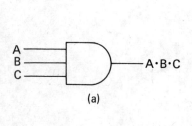

Condition	A	B	C	A·B·C
1	0	0	0	0
2	0	0	1	0
3	0	1	0	0
4	0	1	1	0
5	1	0	0	0
6	1	0	1	0
7	1	1	0	0
8	1	1	1	1

(b)

Figure 172. Three-input AND gate
(a) Logic symbol
(b) Truth table

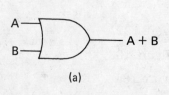

Condition	A	B	A + B
1	F	F	F
2	F	T	T
3	T	F	T
4	T	T	T

(b)

A + B = A OR B

Figure 173. Two-input OR gate
(a) Logic symbol
(b) Truth table

DIGITAL LOGIC TRUTH TABLES

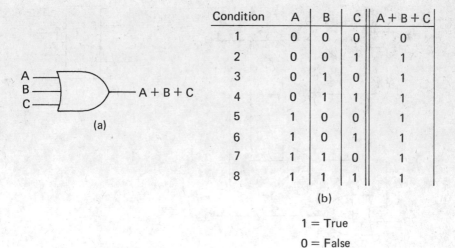

Condition	A	B	C	A + B + C
1	0	0	0	0
2	0	0	1	1
3	0	1	0	1
4	0	1	1	1
5	1	0	0	1
6	1	0	1	1
7	1	1	0	1
8	1	1	1	1

(b)

1 = True
0 = False

Figure 174. Three–input OR gate
(a) Logic symbol
(b) Truth table

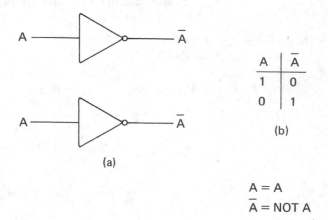

A	$\overline{A}$
1	0
0	1

(b)

(a)

$A = A$
$\overline{A} = $ NOT A

Figure 175. Inverter
(a) Logic symbol
(b) Truth table

175

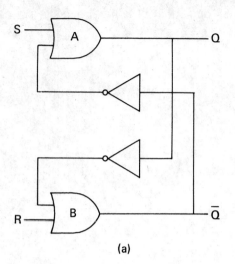

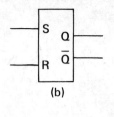

(b)

(a)

	S	R	Q	$\overline{Q}$
First condition	0	0	No change	
Second condition	0	1	0	1
Third condition	1	0	1	0
Fourth condition	1	1	Invalid	

(c)

Figure 176.   RS flip–flop.
(a) Logic diagram
(b) Symbol
(c) Truth table

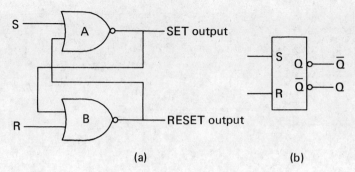

(a)   (b)

Input		Command	Output	
S	R		Q	$\overline{Q}$
L	L	Remember	No change	
L	H	Set	L	H
H	L	Reset	H	L
H	H	Invalid	L	L

(c)

H = Logic high, or True

L = Logic low, or False

Figure 177. RS flip-flop (NOR gate configuration)
(a) Logic diagram
(b) Symbol
(c) Truth table

SECTION 19

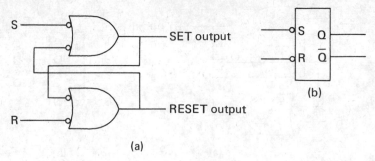

(a)

(b)

Input		Command	Output	
S	R		Q	Q̄
L	L	Invalid	H	H
L	H	Set	H	L
H	L	Reset	L	H
H	H	Remember	No change	

(c)

Figure 178. RS flip-flop (NAND gates using equivalent NOR symbol)
(a) Logic diagram
(b) Symbol
(c) Truth table

178

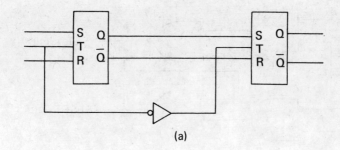

(a)

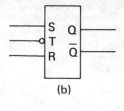

(b)

Input		Command	Output	
S	R		Q	$\overline{Q}$
L	L	Remember	NC	NC
L	H	Reset	L	H
H	L	Set	H	L
H	H	Invalid	?	?

(c)

NC = No Change
? = Indeterminate

Figure 179. Master–slave flip–flop using gated RS flip–flops
(a) Logic diagram
(b) Symbol
(c) Truth table

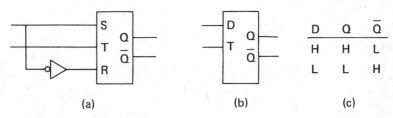

(a)	(b)	(c)

D	Q	$\overline{Q}$
H	H	L
L	L	H

Figure 180. Type D flip–flop
(a) RST flip–flop and inverter connected to form a
Type D flip–flop
(b) Symbol for pre–wired Type–D flip–flop
(c) Truth table

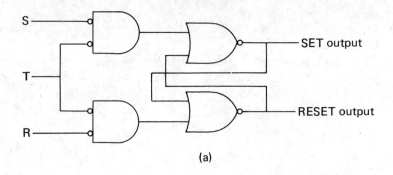

(a)

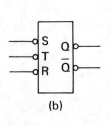

(b)

Input			Command	Output	
S	R	T		Q	Q̄
L	L	L	Invalid	L	L
L	L	H		NC	NC
L	H	L	Set	H	L
L	H	H		NC	NC
H	L	L	Reset	L	H
H	L	H		NC	NC
H	H	L	Remember	NC	NC
H	H	H		NC	NC

(c)

NC = No Change

Figure 181. Gated RS flip–flop
(a) Logic diagram
(b) Symbol
(c) Truth table

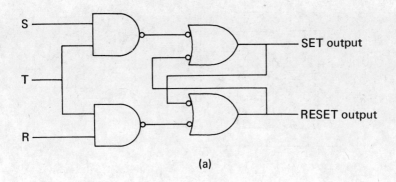

(a)

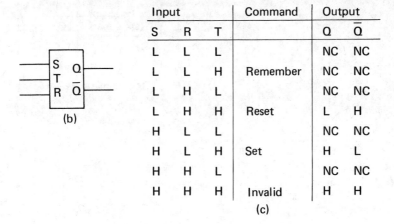

(b)

Input			Command	Output	
S	R	T		Q	$\overline{Q}$
L	L	L		NC	NC
L	L	H	Remember	NC	NC
L	H	L		NC	NC
L	H	H	Reset	L	H
H	L	L		NC	NC
H	L	H	Set	H	L
H	H	L		NC	NC
H	H	H	Invalid	H	H

(c)

NC = No Change

Figure 182. Gated RS flip–flop (NAND gates)
(a) Logic diagram
(b) Symbol
(c) Truth table

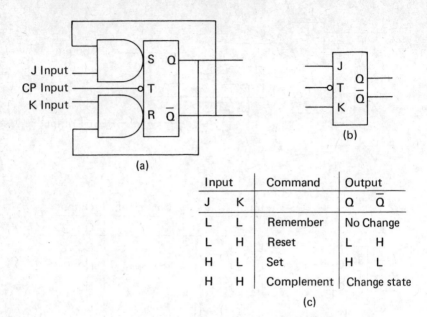

Input		Command	Output	
J	K		Q	Q̄
L	L	Remember	No Change	
L	H	Reset	L	H
H	L	Set	H	L
H	H	Complement	Change state	

(c)

Figure 183. J–K flip–flop (RST flip–flop/gates).
(a) RST flip–flop and two gates connected to form a J–K flip–flop
(b) Symbol
(c) Truth table

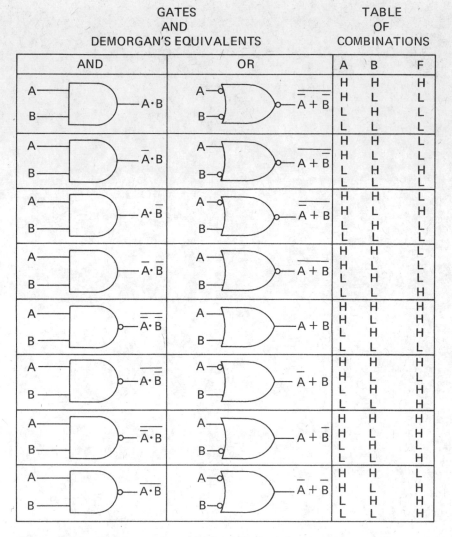

Figure 184. DeMorgan's equivalent basic gates.

SECTION 20

OPERATIONAL AMPLIFIERS

The operational amplifier (op amp) performs computing functions of addition, subtraction, integration, and differentiation, and is used in applications such as signal conditioning, analog instrumentation, active filters, servo systems, process control, nonlinear function generators, regulators and other functions.

The versatility of the op amp results from the use of a large amount of negative feedback around the device. The characteristics of the amplifier in a given application with feedback are determined only by the external feedback elements.

THE BASIC OPERATIONAL AMPLIFIER

The operational amplifier has a single-ended output and differential input (Figure 185). The circuit amplifies the difference between the voltages applied to its two input terminals; a positive voltage at the inverting input produces a negative output. A positive voltage at the non-inverting input produces a positive output.

The characteristics of an ideal op amp are:

1. Infinite voltage gain
2. Infinite input resistance
3. Zero output resistance
4. Infinite bandwidth
5. Zero offset.

Basic properties of an ideal op amp are infinite input resistance and infinite gain. Because of infinite input resistance, *no current flows into either input terminal*. When negative feedback is applied around the amplifier, *the differential input voltage is zero* because of infinite gain. These two basic properties make possible rapid analysis of any operational amplifier circuit.

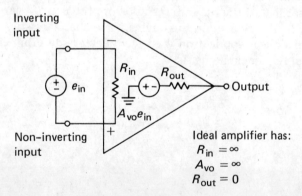

Figure 185. Basic operational amplifier equivalent circuit

INVERTING AMPLIFIER

A basic feedback circuit, the inverting amplifier, is shown in Figure 186. Since the differential input voltage is zero, the potential at the inverting input equals that of the non-inverting input (ground), and the two currents are:

$$I_1 = \frac{V_{in}}{R_s} \qquad\qquad 20.1$$

$$I_2 = \frac{V_{out}}{R_f} \qquad\qquad 20.2$$

No current flows into the amplifier, $I_1 = I_2$, and the voltage gain is

$$A_V = \frac{V_{out}}{V_{in}} = -\frac{R_f}{R_s} \qquad\qquad 20.3$$

The input impedance is R_s, the output impedance is zero. Any value of gain can be obtained by proper choice of the feedback elements. Point P is called the summing point.

The virtual ground at the summing point allows any number of input voltages to be applied to the amplifier and summed at the output without interaction between the sources. Each input sees its respective resistor as the input resistance, and the current in the feedback resistor is the algebraic sum of the current from each input source.

The operational amplifier may be used as a current source by putting the load in the feedback loop (i.e., $R_L = R_f$ in Figure 186.) The current through the load given by Equation 20.1 is completely independent of the load impedance.

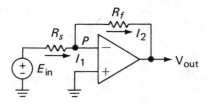

Figure 186. Inverting amplifier

NON-INVERTING AMPLIFIER

The non-inverting amplifier, or potentiometric connection, is shown in Figure 187, page 188. As no current flows into the amplifier inputs, R_f and R_s form a simple voltage divider, and the voltage at the inverting input is equal to

$$V_i = \frac{R_s}{R_f + R_s} V_{out} \qquad\qquad 20.4$$

This voltage is equal to the input voltage because of the infinite voltage gain.

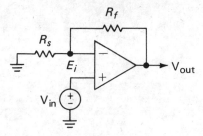

Figure 187. Non-inverting amplifier

$$\frac{V_{out}}{V_{in}} = 1 + \frac{R_f}{R_s} \qquad\qquad 20.5$$

The input impedance is infinite, the output impedance is zero, and only voltage gains above unity are possible.

A unity-gain buffer, or voltage-follower, results when the source resistor (R_s) is removed; any value of R_f can be used in the ideal case since no current flows in the feedback path.

DIFFERENTIAL AMPLIFIER

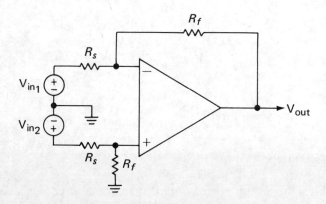

Figure 188. Differential amplifier

The differential amplifier (Figure 188) uses both inputs to the operational amplifier. Consider each input separately. The output voltage due to the signal on the inverting input is

$$V_{out_1} = -\frac{R_f}{R_s} V_{in_1} \qquad\qquad 20.6$$

and the output due to the signal at the non–inverting input is

$$V_{out\,2} = (1 + \frac{R_f}{R_s})\,(\frac{R_f}{R_f + R_s})\,V_{in\,2} \qquad 20.7$$

A voltage divider at the non–inverting input makes the overall gains from both signals equal, so the total output voltage will be

$$V_{out} = \frac{R_f}{R_s}\,(V_{in\,2} - V_{in\,1}) \qquad 20.8$$

FEEDBACK ELEMENTS

Any form of complex or non–linear feedback may be used as feedback elements. The overall transfer function may be derived with appropriate changes being made to include the reactive or non–linear relationships.

Using a capacitor as the feedback element will integrate the input voltage. A differentiator is obtained with the capacitor as the source and the resistor as the feedback element. Logarithmic output results from a diode or transistor in the feedback path. Many other possibilities exist.

ERROR SOURCES

Offset and its variations with temperature, time, supply voltage and common mode voltage is a limitation in dc amplifiers. Due to the component mismatch, the output will have a dc offset when there is no input signal. Offset is independent of the amplifier gain and is compared to the input signal.

An equivalent circuit for the op amp showing the principal sources of offset is given in Figure 189. The input offset voltage (V_{os}) is defined as the voltage that must be applied between the input terminals to obtain zero output. Input bias current is the average of the two currents

$$\frac{(I_{b1} + I_{b2})}{2} \qquad 20.9$$

that flow into the inputs when the output is nulled, and input offset current (I_{os}) is the difference of the two input currents under the same conditions. Both voltage and current offset, together with the impedance levels of the external components, must be considered when predicting the performance of an operational amplifier circuit.

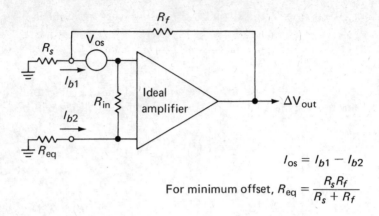

$$I_{os} = I_{b1} - I_{b2}$$

$$\text{For minimum offset, } R_{eq} = \frac{R_s R_f}{R_s + R_f}$$

Figure 189. Equivalent circuit with sources of offset shown

The output offset voltage developed by the basic circuit of Figure 189 is:

$$\Delta \ V_{out} = (1 + \frac{R_f}{R_s}) V_{os} + I_{b1} R_f - I_{b2} R_{eq} (1 + \frac{R_f}{R_s}) \qquad 20.10$$

The two bias currents in a well–matched differential amplifier are approximately equal. Therefore, their effect upon offset can be cancelled by making the impedance to ground equal at both inputs. Therefore:

$$R_{eq} = \frac{R_s R_f}{R_s + R_f}$$

The output offset is then:

$$\Delta \ V_{out} = (1 + \frac{R_f}{R_s}) V_{os} + R_f I_{os} \qquad 20.11$$

In those applications where it is not possible to equalize the resistance, the contribution of the input bias currents must be considered, as shown in Eq. 20.10.

Note that the offset produced at the output is independent of the amplifier being operated in the inverting or non–inverting connection (Eq. 20.11). This means that the offset error will be different for the two configurations, even though the voltage gains are equal. For the inverting amplifier, the input–referred offset is

$$\Delta \ V_{in} = (1 + \frac{R_s}{R_f}) V_{os} + R_s I_{os} \qquad 20.12$$

while for the non–inverting amplifier it is

$$\Delta V_{in} = V_{os} + I_{os} \left(\frac{R_s R_f}{R_f + R_s}\right) \qquad 20.13$$

A fixed-input offset is not usually a problem because biasing circuits can be added to cancel it out. However, drift of offset with temperature, time, and so forth, introduces an input error because this change cannot be distinguished from the input signal.

OPEN-LOOP GAIN

Ideal transfer characteristics cannot be obtained in an op amp because the amplifier gain is not infinite. The accuracy of the circuit is limited to that of the passive components of the circuit.

The gain relationship for a closed–loop circuit, where the amplifier has a finite open-loop gain (A_{vo}) from Figure 185 is

$$A = \frac{1}{1 + \frac{1}{\beta A_{vo}}} \qquad 20.14$$

where

$$\frac{1}{\beta} = \left(1 + \frac{R_f}{R_g}\right)\left(1 + \frac{R_{eq} + \frac{R_s R_f}{R_s + R_f}}{R_{in}}\right) \qquad 20.15$$

The term βA_{vo} is the "loop gain," which determines how close to ideal an amplifier can come.

The percentage of error due to the finite gain of an amplifier is

$$\epsilon\% = \frac{100}{\beta A_{vo}} \qquad 20.16$$

Gain stability is found by the formula

$$\frac{\Delta A}{A} = \frac{\frac{A_{vo}}{A_{vo}}}{1 + \beta A_{vo}} \qquad 20.17$$

Generally, gain stability, distortion and linearity can be improved by the loop gain. Input and output impedance can be changed by the loop gain.

FREQUENCY RESPONSE

Like all amplifying devices, operational amplifiers have the ability and
inclination for oscillation under appropriate feedback conditions. In-
spection of Equation 20.14 reveals that if the loop gain equals unity
and its sign is negative (positive feedback), the circuit will oscillate.
Therefore, it is necessary to ensure that the loop gain be reduced to
less than unity before the loop phase shift reaches 180 degrees. This
is accomplished by designing the loop gain to have a uniform roll-off

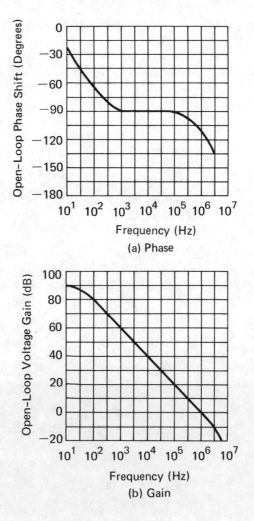

Figure 190. Typical circuit responses
          (a) Phase response
          (b) Frequency response

with frequency, at a rate of about 6dB/octave, beginning at a relatively low frequency and continuing until it passes through unity. Since the phase shift associated with such a roll-off (Figure 190) is 90 degrees, the circuit cannot become unstable.

The closed-loop response of an operational amplifier to a pulse or step function input depends upon the amplitude of the signal. For small signals, the output will be an exponential with a time constant inversely proportional to the closed-loop bandwidth. The closed-loop bandwidth, as seen above, can be maximized for any particular gain and does not necessarily have to be inversely proportional to the closed-loop gain, as with discrete amplifiers.

Apart from bandwidth, operational amplifiers have limitations on the maximum rate of change that the output can follow for large input signals. This is a result of the finite current available to charge the internal capacitances at various nodes within the amplifier. The maximum rate of change, or slew rate, also defines the maximum frequency where full sine-wave output swing can be obtained from the amplifier. The slew rate (ρ) is related to the peak sine-wave signal (A_p) at a given frequency (ω) by the equation

$$\rho = A_p \omega \qquad\qquad 20.18$$

Driving an amplifier beyond its slewing limit results in a triangular output that decreases with increasing frequency. It can also disturb the dc conditions within the amplifier and cause an effective change in the offset voltage, as mentioned previously.

NOISE

Any spurious signal at the output of an amplifier that is not present in the input signal can be considered as noise. For ac amplifiers, random noise generated with the amplifier limits the smallest signal that can be distinguished at the input, while in dc applications, offset voltage and offset drift are the dominant factors. Like offset and drift, random noise can be characterized by a series noise-voltage generator and a parallel noise-current generator at the amplifier inputs, as shown in Figure 191, page 194.

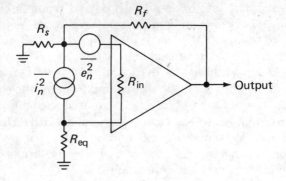

Figure 191. Operational amplifier with noise sources shown.

APPENDIX

This appendix contains multiple and submultiple units, standard letter symbols, wire table, characteristics of radio waves, single-layer coil calculator, resistor color code, capacitor color code, transformer color codes, waveguide symbols, and a glossary of semiconductor terms.

Figure 192. MULTIPLE AND SUBMULTIPLE UNITS

Multiply Reading In	By	To Obtain Reading In
Amperes	1 000 000 000 000	Micromicroamperes
Amperes	1 000 000	Microamperes
Amperes	1000	Milliamperes
Farads	1 000 000 000 000	Micromicrofarads
Farads	1 000 000	Microfarads
Farads	1000	Millifarads
Henries	1 000 000	Microhenries
Henries	1000	Millihenries
Volts	1 000 000	Microvolts
Volts	1000	Millivolts
Mhos	1 000 000	Micromhos
Mhos	1000	Millimhos
Watts	1000	Milliwatts
Hertz	0.000 001	Megahertz per second
Hertz	0.001	Kilohertz per second
Microamperes	0.000 001	Amperes
Milliamperes	0.001	Amperes
Micromicrofarads } Picofarads	0.000 000 000 001	Farads
Microfarads	0.000 001	Farads
Millifarads	0.001	Farads
Microhenries	0.000 001	Henries
Microvolts	0.000 001	Volts
Millivolts	0.001	Volts
Micromhos	0.000 001	Mhos
Millimhos	0.001	Mhos
Milliwatts	0.001	Watts
Kilowatts	1000	Watts
Megahertz	1 000 000	Hertz
Kilohertz	1000	Hertz
Megohms	1 000 000	Ohms

Figure 193. STANDARD LETTER SYMBOLS

B	=	Susceptance
C	=	Capacitance
D	=	Total harmonic distortion
E, e	=	Electromotive force
F_v	=	Power factor
f	=	Frequency
g, G	=	Conductance
H_1	=	Fundamental frequency component of distortion
H_2, H_3, etc.	=	Second (third, etc.) harmonic components of distortion
I, i	=	Current
K	=	Dielectric constant
L	=	Inductance
M	=	Mutual inductance
P	=	Power
Q	=	Charge, quantity of electricity
also Q	=	Figure of merit of a reactor
R	=	Resistance
X	=	Reactance
X_L	=	Inductive reactance
X_c	=	Capacitive reactance
Y	=	Admittance
Z	=	Impedance (scalar)
$\mathbf{Z}$	=	Impedance (vector)

Figure 194. WIRE TABLE

B & S Gage	Bare Diameter	Diameter of Insulated Wire									Area, cir mils	Ohms per 1,000 ft at 25°C	Ft per ohm at 25°C	Lb per 1,000 ft
		Single Enamel	Double Enamel	Single Cotton Enamel	Single Silk Enamel	Single Silk	Single Cotton	Double Cotton	Single Silk	Double Silk				
44	.0020	.0023									4.00	2,700	.3850	.012
43	.0022	.0025									4.84	2,150	.4670	.015
42	.0025	.0029									6.25	1,700	.6050	.019
41	.0028	.0032									7.84	1,350	.7630	.024
40	.0031	.0036	.0039								9.61	1,103	.9550	.030
39	.0035	.0040	.0044								12.25	864	1.204	.038
38	.0040	.0046	.0050								16.00	659	1.519	.048
37	.0045	.0051	.0055								20.30	522	1.915	.060
36	.0050	.0057	.0061	.0095	.0075	.0090	.0130	.0090	.0070	.0090	25.00	424	2.414	.076
35	.0056	.0064	.0067	.0102	.0082	.0096	.0136	.0096	.0076	.0096	31.40	338	3.045	.096
34	.0063	.0072	.0077	.0109	.0089	.0103	.0143	.0103	.0083	.0103	39.70	266	3.839	.120
33	.0071	.0080	.0085	.0117	.0097	.0111	.0151	.0111	.0091	.0111	50.40	210	4.841	.152
32	.0080	.0090	.0095	.0127	.0107	.0120	.0160	.0120	.0100	.0120	64.00	165	6.105	.19
31	.0089	.0100	.0104	.0137	.0117	.0129	.0169	.0129	.0109	.0129	79.20	134	7.698	.24
30	.0100	.0111	.0117	.0148	.0128	.0140	.0180	.0140	.0120	.0140	100	106	9.707	.31

29	.0113	.0125	.0130	.0162	.0142	.0153	.0193	.0133	.0153	128	83.1	12.24	.38
28	.0126	.0139	.0145	.0175	.0155	.0166	.0206	.0146	.0166	159	66.4	15.43	.48
27	.0142	.0155	.0161	.0192	.0172	.0182	.0222	.0162	.0182	202	52.5	19.46	.61
26	.0159	.0172	.0178	.0210	.0190	.0199	.0239	.0179	.0199	253	41.7	24.54	.77
25	.0179	.0193	.0200	.0234	.0211	.0222	.0262	.0199	.0219	320	33.0	30.95	.97
24	.0201	.0216	.0222	.0256	.0233	.0244	.0284	.0221	.0241	404	26.2	39.02	1.23
23	.0226	.0242	.0247	.0282	.0259	.0269	.0309	.0246	.0266	511	20.7	49.21	1.54
22	.0253	.0271	.0278	.0310	.0287	.0296	.0336	.0273	.0293	645	16.4	62.05	1.95
21	.0285	.0302	.0310	.0344	.0319	.0330	.0370	.0305	.0325	812	13.0	78.25	2.45
20	.0320	.034	.0345	.0385	.0355	.0370	.0410	.0340	.0360	1,020	10.3	98.66	3.09
19	.0359	.038	.0387	.0425	.0395	.0409	.0449	.0379	.0399	1,300	8.14	124.4	3.89
18	.0403	.042	.0431	.0469	.0439	.0453	.0493	.0423	.0443	1,600	6.59	156.9	4.9
17	.0453	.047	.0481	.0521	.0491	.0503	.0543	.0473	.0493	2,030	5.22	197.8	6.2
16	.0508	.053	.0536	.0576	.0546	.0558	.0608	.0528	.0548	2,600	4.07	249.4	7.8
15	.0571	.059	.0605	.0640	.0610	.0621	.0671	.0591	.0611	3,250	3.26	314.5	9.94
14	.0641	.066	.0675	.0711	.0681	.0691	.0741	.0661	.0681	4,100	2.58	396.6	12.4
13	.0719									5,180	2.00	499.3	15.7
12	.0808									6,530	1.59	629.6	19.8
11	.0907									8,235	1.26	794.0	24.9
10	.1019									10,380	1.00	1,001	31.4
9	.1144									13,090	.792	1,262	40.0
8	.1285									16,510	.628	1,592	50.0

Figure 194

Figure 195. CHARACTERISTICS OF RADIO WAVES

Frequency	3 kHz — 30 kHz	30 kHz — 300 kHz	300 kHz — 3 MHz	3 MHz — 30 MHz	30 MHz — 300 MHz	300 MHz — 3 kMHz	3 kMHz — 30 kMHz	30 kMHz — 300 kMHz
Designation	VLF	LF	MF	HF	VHF	UHF	SHF	EHF
Bands						P L S	S X K (Microwaves)	
Maximum range (miles)	Worldwide	3000	5000	12,000	Line of sight (sporadic long ranges)	Line of sight	Line of sight	Line of sight
Propagation	D-layer reflection	D-layer reflection	E-layer reflection	F-layer reflection	Sporadic ionosphere reflection	Sporadic atmosphere reflection	Sporadic ducting and some atmospheric absorption	Great atmospheric absorption
External interference sources	Man-made and natural	Man-made and natural	Man-made and natural	Man-made and natural	Man-made	None	None	None
Transmission lines	Open wire	Open wire	Open wire	Open wire	Open wire and coaxial	Coaxial, open wire and waveguide	Waveguide	Waveguide
Applications	Communications, experimental	Communications, navigation	Communications, navigation	Communications, navigation, control, medical	Communications, navigation, television, control, relay, radar, industrial, medical	Communications, navigation, television, control, relay, radar, medical	Communications, navigation, control, relay, radar, industrial, nuclear resonance	Atomic, clocks, nuclear resonance, radar, control, experimental

Figure 196. SINGLE-LAYER COIL CALCULATOR

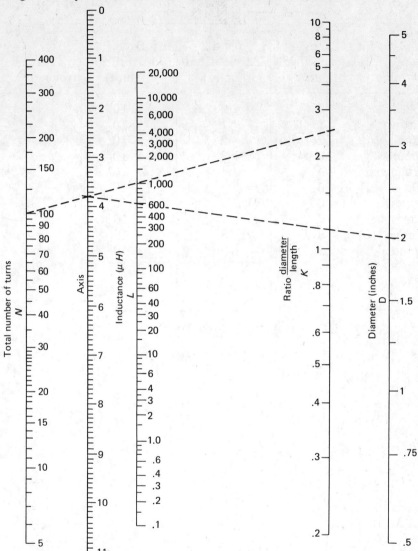

Courtesy: Sprague Products Inc.

To find the number of turns required, proceed as follows:
1. Connect 290 on L to 2.5 on D, and note point where the line crosses the axis (4.6).
2. Connect 4.6 on axis to 0.8 on K.
3. Read number of turns where line crosses N (90 turns).

APPENDIX

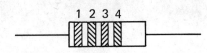

Color	Band 1 (1st significant figure)	Band 2 (2d significant figure)	Band 3 (3d significant figure)	Band 4 (tolerance, %)
Black	0	0	$\times 10^0$	-
Brown	1	1	$\times 10^1$	-
Red	2	2	$\times 10^2$	-
Orange	3	3	$\times 10^3$	-
Yellow	4	4	$\times 10^4$	-
Green	5	5	$\times 10^5$	-
Blue	6	6	$\times 10^6$	-
Purple	7	7	$\times 10^7$	-
Gray	8	8	$\times 10^8$	-
White	9	9	$\times 10^9$	-
No color	-	-	-	20
Silver	-	-	-	10
Gold	-	-	-	5

Figure 197. RESISTOR COLOR CODE

Figure 198. CAPACITOR COLOR CODE

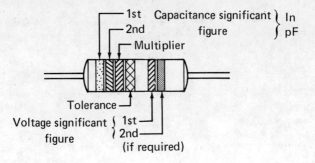

Color Coding of Molded Tubular Capacitors

Color	Significant Figure	Decimal Multiplier	Tolerance ±%
Black	0	1	20
Brown	1	10	-
Red	2	100	-
Orange	3	1000	30
Yellow	4	10,000	40
Green	5	10^5	5
Blue	6	10^6	-
Violet	7	-	-
Gray	8	-	-
White	9	-	10

Note: Voltage rating is identified by a single-digit number for ratings up to 900V; a two-digit number above 900V. Two zeros follow the voltage figure. Courtesy: Sprague Electric Company.

Figure 199.   TRANSFORMER COLOR CODES

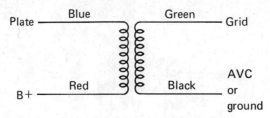

(a) Color code for IF transformers.

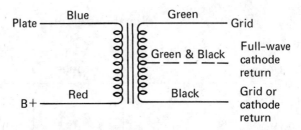

(b) Color code for interstage audio transformers.

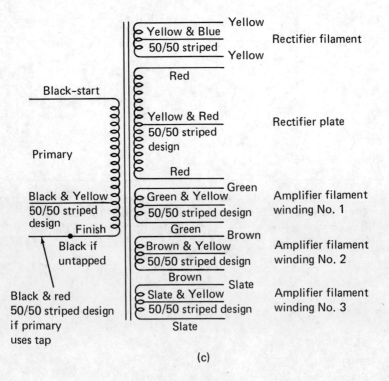

(c)

APPENDIX

Standard colors used in chassis wiring
for the purpose of circuit identification of
the equipment are as follows:

Circuit	Color
Grounds, grounded elements, and returns	Black
Heaters or filaments, off ground	Brown
Power supply B plus	Red
Screen grids	Orange
Cathodes	Yellow
Control grids	Green
Plates	Blue
Power supply, minus	Violet (purple)
ac power lines	Gray
Miscellaneous, above or below ground returns, AVC, etc.	White

(d)

Figure 200. WAVEGUIDE SYMBOLS

Description	Symbol*	Description	Symbol*
Capacitive reactance	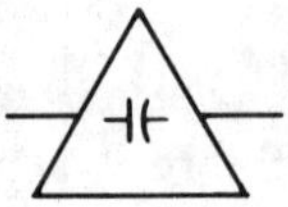	Parallel resonant circuit—zero susceptance	
Capacitive susceptance	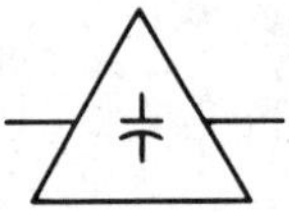	Series resonant circuit—zero reactance	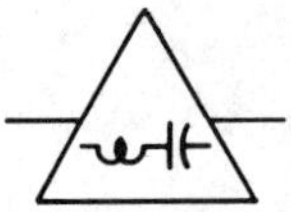
Inductive reactance		Series resonant circuit—infinite susceptance	
Inductive susceptance	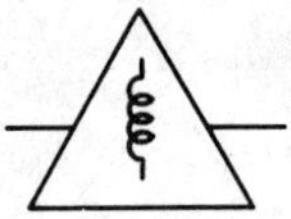	Attenuator	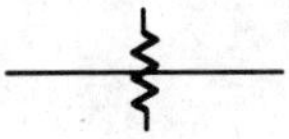
Resistance	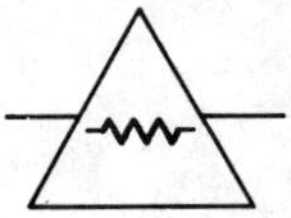	Phase shifter	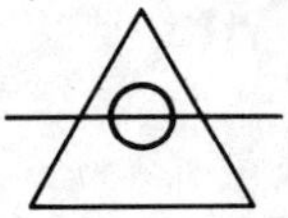
Conductance		Coaxial cable	
Parallel resonant circuit—infinite reactance		Circular waveguide	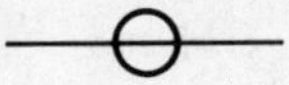

WAVEGUIDE SYMBOLS

Description	Symbol*	Description	Symbol*
Rectangular waveguide		Directional coupler with resistance coupling, 20 dB attenuation	20 dB
Waveguide load and shorted termination		Mode suppression	
Loop coupling		Mode transducer	
Waveguide-to-coaxial-cable coupling (transition)		Rectangular-to-circular-waveguide transducer	
Types of aperture coupling	E H EH		
Directional coupler			
Directional coupler with E type aperture coupling, 20 db attenuation	E 20 dB		

*An arrow drawn diagonally through any symbol indicates that the circuit element it represents is variable.

Figure 201. TRANSFORMER COLOR CODE (RETMA STANDARD)

Color of Lead	Power Transformer	AF Transformer (Also line-to-grid and tube-to-line)	IF Transformer	Loudspeaker Field Coil	Loudspeaker Voice Coil
Black	Primary (common for tapped primary)	Grid return	Grid or diode return	-	Start
Black and red	Finish or tapped primary	-	-	Start	-
Black and yellow	Primary	-	-	-	-
Red	High voltage	B^+	B^+	-	-
Red and yellow	High-voltage tap	-	-	Finish	-
Yellow	Rectifier filament (C.T.—yellow and blue)	Grid or center-tapped secondary	-	-	-
Green	Filament No. 1 (C.T.—green and yellow)	Grid	Grid or diode	-	Finish
Brown	Filament No. 2 (C.T.—brown and yellow)	Plate or center-tapped primary	-	-	-
Slate	Filament No. 3 (C.T.—slate and yellow)	-	-	Tap—slate and red	-
Blue		Plate	Plate	-	-

Figure 202. CERAMIC CAPACITOR COLOR CODE

Color	Significant Figure	Multiplier	Tolerance A	Tolerance B	Temperature Coefficient
Black	0	1	2	20	0
Brown	1	10	.1	1	−30
Red	2	10^2	-	2	−60
Orange	3	10^3	-	2.5	−150
Yellow	4	10^4	-	-	−220
Green	5	-	5	5	−330
Blue	6	-	-	-	−470
Violet	7	-	-	-	−750
Gray	8	.01	.25	-	+30
White	9	.1	1	10	+120 to −750 (RETMA) / +500 to −330 (JAN)
Gold	-	-	-	-	+100
Silver	-	-	-	-	Bipass or coupling

Feed through type

Multiplier — 1st, 2nd Significant figure — Tolerance — Temperature coefficient

Standoff type

1st, 2nd Significant figure — Multiplier — Temperature coefficient — Tolerance

Disc type
5-dot system

3-dot system

Multiplier — 1st, 2nd Significant figure

Temperature coefficient — 1st, 2nd Significant figure — Multiplier — Tolerance

Radial lead type
5-dot system

2nd, 1st Significant figure — Tolerance — Multiplier — Temperature coefficient

6-dot system

1st, 2nd Significant figure — Multiplier — Silver (means bypass or coupling only) — Tolerance — Voltage

Brown–150
Orange–350
Green–500
None–500

Axial lead

Temperature coefficient — 1st, 2nd Significant figure — Multiplier — Tolerance

Figure 203. MICA-CAPACITOR COLOR CODE

Button silver

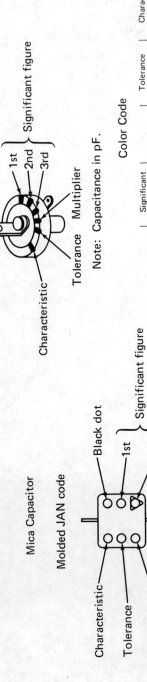

Mica Capacitor

Molded JAN code

Characteristic
Tolerance
Multiplier

Black dot
1st · 2nd } Significant figure

Retma code

Characteristic
Tolerance
Multiplier

White dot
1st · 2nd } Significant figure

Characteristic
Tolerance Multiplier

1st
2nd
3rd } Significant figure

Note: Capacitance in pF.

Note: Capacitance in pF. If both rows of dots are not on one face, rotate capacitor about the axis of its leads and read second row on side or rear.

Color Code

Color	Significant Figure	Multiplier	Tolerance ± %	Characteristic
Black	0	1	20	A
Brown	1	10	-	B
Red	2	10^2	2	C
Orange	3	10^3	3 (RETMA)	D
Yellow	4	10^4	-	E
Green	5	-	5 (RETMA)	F (JAN)
Blue	6	-	-	G (JAN)
Violet	7	-	-	-
Gray	8	-	-	I (RETMA)
White	9	-	-	J (RETMA)
Gold	-	.1	5 (JAN)	-
Silver	-	.01	10	-
None	-	-	20 (RETMA)	-

Figure 204. PAPER CAPACITOR COLOR CODE

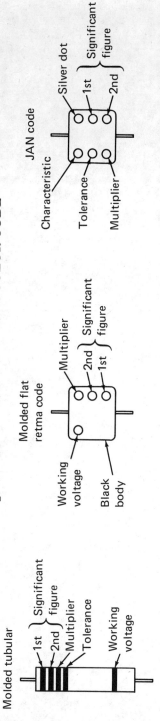

Molded tubular

- 1st, 2nd } Significant figure
- Multiplier
- Tolerance
- Working voltage

Molded flat retma code

- Multiplier
- 2nd, 1st } Significant figure
- Working voltage
- Black body

JAN code

- Silver dot
- 1st, 2nd } Significant figure
- Characteristic
- Tolerance
- Multiplier

Color	Significant Figure	Multiplier	Tolerance ± %
Black	0	1	20
Brown	1	10	-
Red	2	10^2	-
Orange	3	10^3	30
Yellow	4	10^4	40
Green	5	10^5	5
Blue	6	10^6	-
Violet	7	-	-
Gray	8	-	-
White	9	-	10
Gold	-	1	-

Note: Capacitance in picofarads. Working voltage coded in terms of hundreds of volts. Ratings over 900V expressed in two-band voltage code.

Color	Significant Figure	Multiplier	Tolerance ± %	Characteristic
Black	0	1	20	A
Brown	1	10	-	B
Red	2	10^2	-	C
Orange	3	10^3	30	D
Yellow	4	10^4	40	E
Green	5	10^5	5	F
Blue	6	10^6	-	G
Violet	7	-	-	-
Gray	8	-	-	-
White	9	-	10	-
Gold	-	1	5	-

Note: Capacitance in picofarads. Working voltage coded in terms of hundreds of volts. Voltage ratings over 900V expressed in two-dot voltage code.

GLOSSARY OF SEMICONDUCTOR TERMS *

*Courtesy, U.S. Armed Forces.

Acceptor Impurity—A substance with three (3) electrons in the outer orbit of its atom that, when added to a semiconductor crystal, provides one hole in the lattice structure of the crystal.

Amplifier, Class A—An amplifier in which the swing of the input signal is always on the linear portion of the characteristic curves of the amplifying device.

Amplifier, Class AB—An amplifier that has the collector current or voltage at zero for less than half of a cycle of input signal.

Amplifier, Class B—An amplifier that operates at collector current cutoff or at zero collector voltage and remains in this condition for one-half cycle of the input signal.

Amplifier, Class C—An amplifier in which the collector voltage or current is zero for more than a half cycle of the input signal.

AND Circuit (AND Gate)—A coincidence circuit that functions as a gate so that when all the inputs are applied simultaneously, a prescribed output condition exists.

AND-OR Circuit (AND-OR Gate)—A gating circuit that produces a prescribed output condition when several possible combined input signals are applied; exhibits the characteristics of the *AND* gate and the *OR* gate.

Astable Multivibrator—A multivibrator that can function in either of two semistable states, switching rapidly from one to the other; referred to as free running.

Barrier—In a semiconductor, the electric field between the acceptor ions and the donor ions at a junction. (See *Depletion Region*.)

Barrier Height—In a semiconductor, the difference in potential from one side of a barrier to the other.

Base (junction transistor)—The center semiconductor region of a double junction (*NPN* or *PNP*) transistor. The base is comparable to the grid of an electron tube.

Base Spreading Resistance—In a transistor, the resistance of the base region caused by the resistance of the bulk material of the base region.

Beat Frequency Oscillator—An oscillator that produces a signal that mixes with another signal to provide frequencies equal to the sum and difference of the combined frequencies.

Bipolar Transistor—A junction-type transistor employing both majority and minority carriers.

Bistable Multivibrator—A circuit with two stable states requiring two input pulses to complete a cycle.

Blocking Oscillator—A relaxation-type oscillator that conducts for a short period of time and is cut off for a relatively long period of time.

Clamping Circuit—A circuit that maintains either or both amplitude extremities of a waveform at a certain level or potential.

214

GLOSSARY OF SEMICONDUCTOR TERMS

Collector—The end semiconductor material of a double junction (*NPN* or *PNP*) transistor that is normally reverse biased with respect to the base. The collector is comparable to the plate of an electron tube.

Common-Base (CB) *Amplifier*—A transistor amplifier in which the base element is common to the input and the output circuit. This configuration is comparable to the grounded-grid triode electron tube.

Common-Collector (CC) *Amplifier*—A transistor amplifier in which the collector element is common to the input and the output circuit. This configuration is comparable to the electron-tube cathode follower.

Common-Emitter (CE) *Amplifier*—A transistor amplifier in which the emitter element is common to the input and the output circuit. This configuration is comparable to the conventional electron-tube amplifier.

Complementary Symmetry Circuit—An arrangement of *PNP*- and *NPN*-type transistors that provides push-pull operation from one input signal.

Compound-Connected Transistor—A combination of two transistors to increase the current amplification factor at high emitter currents. This combination is generally employed in power amplifier circuits.

Configuration—The relative arrangement of parts (or components) in a circuit.

Constant-Power Dissipation Line—A line (superimposed on the output static characteristic curves) representing the points of collector voltage and current, the products of which represent the maximum collector power rating of a particular transistor.

Crossover Distortion—Distortion that occurs at the points of operation in a push-pull amplifier where the input signals cross (go through) the zero reference points.

Current Stability Factor—In a transistor, the ratio of a change in emitter current to a change in reverse-bias current flow between the collector and the base.

Cutoff Frequency—The frequency at which the gain of an amplifier falls below 0.707 times the maximum gain.

Dependent Variable—In a transistor, one of four variable currents and voltages that is arbitrarily chosen and considered to vary in accordance with other currents and voltages (independent variable).

Depletion Region (*or Layer*)—The region in a semiconductor containing the uncompensated acceptor and donor ions; also referred to as the space-charge region or barrier region.

Differentiating Circuit—A circuit that produces an output voltage proportional to the rate of change of the input voltage.

Donor Impurity—A substance with electrons in the outer orbit of its atom; added to a

215

semiconductor crystal, it provides one free electron.

Double-Junction Photosensitive Semiconductor—Three layers of semiconductor material with an electrode connection to each end layer. Light energy is used to control current flow.

Dynamic Transfer Characteristic Curve—In transistors, a curve that shows the variation of output current (dependent variable) with variation of input current under load conditions.

Electron-Pair Bond—A valence bond formed by two electrons, one from each of two adjacent atoms.

Elemental Charge—The electrical charge on a single electron (negatron or positron).

Emitter-Follower Amplifier—See *Common-Collector Amplifier.*

Emitter (junction transistor)—The end semiconductor material of a double junction (*PNP* or *NPN*) transistor that is forward biased with respect to the base. The emitter is comparable to the cathode of an electron tube.

Equivalent Circuit—A diagrammatic circuit representation of any device exhibiting two or more electrical parameters.

Fall Time—The length of time during which the amplitude of a pulse is decreasing from 90 percent to 10 percent of its maximum value.

FET—Field-effect transistor.

Forward Bias—In a transistor, an external potential applied to a *PN* junction so that the depletion region is narrowed and relatively high current flows through the junction.

Forward-Short-Circuit, Current-Amplification Factor—In a transistor, the ratio of incremental values of output to input current when the output circuit is ac short-circuited.

Four-Layer Diode—A diode constructed of semiconductor materials resulting in three *PN* junctions. Electrode connections are made to each end layer.

Gating Circuit—A circuit operating as a switch, making use of a short or open circuit to apply or eliminate a signal.

Grounded Base Amplifier—See *Common-Base Amplifier.*

Hole—A mobile vacancy in the electronic valence structure of a semicondcutor. The hole acts similarly to a positive electronic charge having a positive mass.

Hybrid Parameter—The parameters of an equivalent circuit of a transistor that are the result of selecting the input current and the output voltage as independent variables.

Increment—A small change in value.

Independent Variable—In a transistor, one of several voltages and currents chosen arbitrarily and considered to vary independently.

Inhibition Gate—A gate circuit used as a switch and placed in parallel with the circuit it is controlling.

Interelement Capacitance—The capacitance caused by the *PN* junctions between the regions

216

of a transistor; measured between the external leads of the transistor.

JFET—Junction-gate field-effect transistor.

Junction Transistor—A device having three alternate sections of *P*-type or *N*-type semiconductor material. See *PNP Transistor* and *NPN Transistor*.

Lattice Structure—In a crystal a stable arrangement of atoms and their electron-pair bonds.

Majority Carriers—The holes or free electrons in *P*-type or *N*-type semiconductors, respectively.

Minority Carriers—The holes or excess electrons found in the *N*-type or *P*-type semiconductors, respectively.

Monostable Multivibrator—A multivibrator having one stable and one semistable condition. A trigger is used to drive the unit into the semistable state where it remains for a predetermined time before returning to the stable condition.

MOSFET—Metallic-oxide semiconductor field-effect transistor.

Multivibrator—A type of relaxation oscillator for the generation of nonsinusoidal waves in which the output of each of two stages is coupled to the input of the other to sustain oscillations. See *Astable Multivibrator*, *Bistable Multivibrator*, and *Monostable Multivibrator*.

Neutralization—The prevention of oscillation of an amplifier by canceling possible changes in the reactive component of the input circuit caused by positive feedback.

NOR Circuit—An *OR* gating circuit that provides pulse-phase inversion.

NOT AND Circuit—An *AND* gating circuit that provides pulse-phase inversion.

NPN Transistor—A device consisting of a *P*-type section and two *N*-type sections of semiconductor material with the *P*-type in the center.

N-Type Semiconductor—A semiconductor into which a donor impurity has been introduced. It contains free electrons.

Open Circuit Parameters—The parameters of an equivalent circuit of a transistor that are the result of selecting the input current and output current as independent variables.

OR Circuit (OR Gate)—A gate circuit that produces the desired output with only one of several possible input signals applied.

Parameter—A derived or measured value that conveniently expresses performances; for use in calculations.

Photosensitive Semiconductor—A semiconductor material in which light energy controls current-carrier movement.

PN Junction—The area of contact between *N*-type and *P*-type semiconductor materials.

PNP Transistor—A device consisting of an *N*-type section and two *P*-type sections of semiconductor material with the *N*-type in the center.

217

Point-Contact—In transistors, a physical connection made by a metallic wire on the surface of a semiconductor.

Polycrystalline Structure—The granular structure of crystals, which are nonuniform in shape and irregularly arranged.

Preamplifier—A low-level stage of amplification, usually following a transducer.

P-Type Semiconductor—A semiconductor crystal into which an acceptor impurity has been introduced. It provides holes in the crystal lattice structure.

Pulse Amplifier—A wideband amplifier used to amplify square waveforms.

Pulse Repetition Frequency—The number of nonsinusoidal cycles (square waves) that occur in one second.

Pulse Time—The length of time a pulse remains at its maximum value.

Quiescence—The operating condition that exists in a circuit when no input signal is applied to the circuit.

Reverse Bias—An external potential applied to a *PN* junction such as to widen the depletion region and prevent the movement of majority current carriers.

Reverse-Open-Circuit, Voltage-Amplification Factor—In a transistor, the ratio of incremental values of input voltage to output voltage measured with the input ac open-circuited.

Rise Time—The length of time during which the leading edge of a pulse increases from 10 percent to 90 percent of its maximum value.

Saturation (Leakage) Current—The current flow between the base and collector or between the emitter and collector, measured with the emitter lead or the base lead, respectively, open.

Semiconductor — A conductor whose resistivity is between that of metals and insulators in which electrical charge carrier concentration increases with increasing temperature over a specific temperature range.

Short-Circuit Parameters—The parameters of an equivalent circuit of a transistor that are the result of selecting the input and output voltages as independent variables.

Single-Junction Photosensitive Semiconductor—Two layers of semiconductor materials with an electrode connection to each material. Light energy controls current flow.

Spacistor—A semiconductor device consisting of one *PN* junction and four electrode connections characterized by a low transient time for carriers to flow from the input element to the output element.

Stabilization—The reduction of variables in voltage or current not due to prescribed conditions.

Storage Time—The time during which the output current or voltage of a pulse is falling from maximum to zero after the in-

218

put current or voltage is removed.

Stray Capacitance—The capacitance introduced into a circuit by the leads and wires used to connect circuit components.

Surge Voltage (or Current)—A large, sudden change of voltage (or current), usually caused by the collapsing of a magnetic field or the shorting or opening of circuit elements.

Swamping Resistor—In transistor circuits, a resistor placed in the emitter lead to mask (or minimize the effects of) variations in emitter-base junction resistance caused by variations in temperature.

Tetrode Transistor—A junction transistor with two electrode connections to the base (one to the emitter and one to the collector) to reduce the interelement capacitance.

Thermal Agitation—In a semiconductor, the random movement of holes and electrons within a crystal due to the thermal (heat) energy.

Transducer—A device that converts one type of power to another, such as acoustical power to electrical power.

Transistor—A semiconductor device capable of transferring a signal from one circuit to another and producing amplification. See *Junction Transistor*.

Triggered Circuit—A circuit that requires an input signal (trigger) to produce a desired output determined by the characteristics of the circuit.

Trigger Pulse Steering—In transistors, the routing or directing of trigger signals (usually pulses) through diodes or transistors (called steering diodes or steering transistors) so that the trigger signals affect only one circuit of several associated circuits.

Tuned-Base Oscillator—A transistor oscillator with the frequency-determining device (resonant circuit) located in the base circuit. It is comparable to the tuned-grid electron-tube oscillator.

Tuned-Collector Oscillator—A transistor oscillator with the frequency-determining device located in the collector circuit. It is comparable to the tuned-plate electron-tube oscillator.

Turnoff Time—The time that it takes a switching circuit (gate) to completely stop the flow of current in the circuit it is controlling.

Unijunction Transistor — A *PN* junction transistor with one electrode connection to one of the semiconductor materials and two connections to the other semiconductor material.

Unilateralization—The process by which an amplifier is prevented from going into oscillation by canceling the resistive and reactive component changes in the input circuit of an amplifier caused by positive feedback.

Unipolar Transistor—A field-effect device employing majority carriers only.

Unit Step Current (or Voltage)—A

current (or voltage) that undergoes an instantaneous change in magnitude from one constant level to another.

Voltage Gain—The ratio of incremental values of output voltage to input voltage of an amplifier under load conditions.

Wideband Amplifier—An amplifier capable of passing a wide range of frequencies with equal gain.

Zener Diode—A *PN* junction diode reverse biased into the breakdown region; used for voltage stabilization.

INDEX

Figure 1 Reactance Chart*
(Always obtain approximate value from Figure 2 before using Figure 1.)

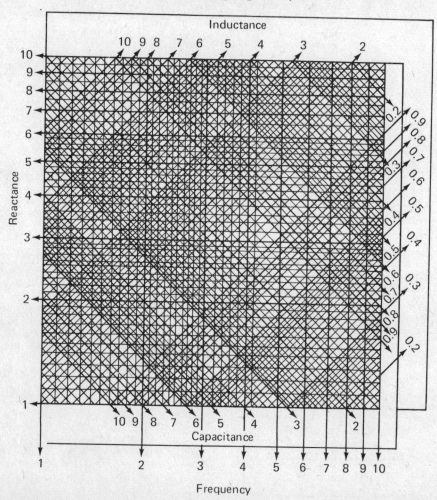

Frequency

Figure 1 is used to obtain greater precision of reading but does not place the decimal point, which must be located from a preliminary entry on Figure 2. Since the chart requires two logarithmic decades for inductance and capacitance for every single decade of frequency and reactance, unless the correct decade for L and C is chosen, erroneous results are obtained.

Typical Results

1. Find the reactance of an inductance of 0.00012 H at 960 kHz.

Answer: 720 Ω.

2. What capacitance will have 265 Ω reactance at 7000 kHz?

Answer: 86 pF.

3. What is the resonant frequency of $L = 21$ μH and $C = 45$ pF?

Answer: 5.18 MHz.

*Figures 1 and 2 are from General Radio Company Reactance Chart.

JAN 1981   HUDSON'S   3.99   Reg 15.95